Thomas Widlok, M. Dores Cruz (eds.)
Scale Matters

AF764549

Culture & Theory | Volume 263

Thomas Widlok is professor for cultural anthropology at the University of Cologne. He completed his MSc and PhD at the London School of Economics and Political Science and was a member of the CRC "Our way to Europe" funded by the German Research Foundation (DFG). He is a specialist in hunter-gatherer studies.
M. Dores Cruz graduated from Binghamton University (SUNY) with a Ph.D. in anthropology. She is currently a lecturer at the Institute for African Studies and Egyptology, Universität zu Köln, and specializes on African historical archaeology, landscape anthropology, and colonialism.

Thomas Widlok, M. Dores Cruz (eds.)

Scale Matters

The Quality of Quantity in Human Culture and Sociality

[transcript]

This publication has been made possible by the generous support of the German Research Foundation (DFG) through the Collaborative Research Centre 806.

Funded by
DFG Deutsche Forschungsgemeinschaft
German Research Foundation

Bibliographic information published by the Deutsche Nationalbibliothek
The Deutsche Nationalbibliothek lists this publication in the Deutsche Nationalbibliografie; detailed bibliographic data are available in the Internet at http://dnb.d-nb.de

This work is licensed under the Creative Commons Attribution-ShareAlike 4.0 (BY-SA) which means that the text may be remixed, build upon and be distributed, provided credit is given to the author and that copies or adaptations of the work are released under the same or similar license. For details go to
http://creativecommons.org/licenses/by-sa/4.0/
Creative Commons license terms for re-use do not apply to any content (such as graphs, figures, photos, excerpts, etc.) not original to the Open Access publication and further permission may be required from the rights holder. The obligation to research and clear permission lies solely with the party re-using the material.

First published in 2022 by transcript Verlag, Bielefeld
© Thomas Widlok, M. Dores Cruz (eds.)

Cover layout: Kordula Röckenhaus, Bielefeld
Printed by Majuskel Medienproduktion GmbH, Wetzlar
Print-ISBN 978-3-8376-6099-9
PDF-ISBN 978-3-8394-6099-3
https://doi.org/10.14361/9783839460993
ISSN of series: 2702-8968
eISSN of series: 2702-8976

Printed on permanent acid-free text paper.

Contents

Introduction: Why scale matters

Thomas Widlok

The issue of scale has been somewhat dormant for a long time in anthropology and archaeology if not in the social sciences and in humanities research more generally. In terms of book publications that tackle scale head-on, the early volume by Fredrik Barth et al. (1978) stands out. It was published in the 1970s at that point in time when globalization received a boost after the Second World War and after many colonies had entered the international scene as independent states. At that stage it had become increasingly difficult in anthropology and its neighbouring disciplines to study people and places as if they had been isolated or as if they had been units that could analytically be demarcated in an unproblematic manner. The immediate result of that awareness, however, was a period of more or less 'peaceful coexistence' between research projects that continued a focus on small-scale case studies (using ethnography) and those who took on the national and international level (using survey and sampling techniques). Only much more recently have anthropologists advocated that the emerging multi-sited ethnography should be complemented also by a multi-scalar ethnography (see Xiang 2013).

In human geography, by contrast, considerable attention has been given to problems of scale in the last few decades. Here, the extensive literature on the topic was, again, triggered by processes of globalization. Globalization is not simply enlarging the scope of social transformations but it provides states and international corporations with the power to implement numerous shifts of scale. For instance, economic differentiation allows development and management to be scaled up in one country or region, where taxation is low, while scaling up manufacture elsewhere, where wage labour is cheap. Scale shifted from being seen as a 'given' dimension to being investigating as something that is part and parcel of political and economic strategies. This led to a situation (at least in human geography) where the social constructedness or 'production' of scale has since become "an established truism" in the discipline (Brenner 2001: 599, see also Marston 2000). Scale is now something that needed explanation rather than providing an explanation for social processes. The naive assumption that scales (manifest in local, regional, national, global 'levels') are unproblematic and nicely nested systems "of territorial containers defined by absolute geographic size" has been discredited (Brenner 2001: 606). The corresponding "'Russian dolls' model of scales" (ibid.) has as a consequence been rigorously undermined even though it continues to live on outside specialized academia. Scale now popped up everywhere and in kaleidoscopic fashion. So much so that it is debated whether notions of 'scale' (in particular 'the politics of scale') have not become hopelessly overstretched as they were extended to cover many other aspects of spatiality that had previously been discussed using other terminologies (Brenner 2001: 596, see Marston 2000 and Marston and Smith 2001).

The main lesson that anthropology and archaeology can learn from these debates in geography is that scale is currently being used in terms of at least three aspects, namely size, level and relation (see Howitt 1998). 'Size' is probably the most common analogue to scale, as illustrated by the classificatory distinction between 'large-scale' and 'small-scale' *societies* (or similar social entities 'out there') as also discussed in contributions to this volume. 'Larger' here does not necessarily refer to a single type of 'size' but could be conceived of in terms of demography, spatiality, institutional complexity or a combination of these.

The second aspect, that of 'level', is also referred to repeatedly in the chapters of this book when contributors discuss whether and how units are 'nested' at multiple levels, e.g. whether hunter-gatherer residential groups can be seen as being recruited along the hierarchically 'nested' levels of camps, bands, 'tribes' or societies. The point here is that we are always dealing with levels as multiple scales, also when dealing with groups of small size, so that any small number of hunter-gatherers staying together can be conceived of as part of a residential group and – at the same time – of larger territorial groups which in turn can be part of larger networks of trade or cultural exchange. 'Levels' can be constructed on the basis of a variety of parameters. Accordingly, 'small-scale', 'large-scale' or 'cross-scale' *studies* can, in this understanding, be conducted in parallel or as complements to one another.

'Relation', the third aspect of scale, is now increasingly taking centre stage, having been undertheorized for a long time (Howitt 1998: 53). Simply put, it recognizes that every unit can be placed in relation to a host of other units or more generally to a number of *situations* (see Widlok 2016). For instance, outside anthropology, cultural features are commonly assessed in relation to 'nations' or 'nation states' while this is rather uncommon in our discipline which is critical of 'methodological nationalism' and which often also has reservations against 'methodological individualism'. Although the measures for 'size' and 'level' are also not given but are constructed and standardized in research, the contingency of scale depending on whoever is doing the scaling is most clearly visible when it comes to the 'relation' aspect of scale. In geography debates underline that the meaning of scalar terms (global, national, regional and urban) "will differ qualitatively depending on the historically and contextually specific scalar partitionings of the sociospatial process in question" (Brenner 2001: 606). Within any particular historical context, scale-making as an embodied practice will inevitably vary since the social agents who take part in these processes are "themselves shaped by gender, race, class and geography" (Marston and Smith 2001: 617).

In contrast to human geography, anthropology and archaeology have been eclectic in their theoretical orientations and interpretations concerning scale. Many archaeologists share the 'materialist' bias that we also find in the human geography debates (following the work of Henri Lefebvre), while being less concerned about the subtleties of concepts surrounding geographical scope. Many socio-cultural anthropologists have limited themselves to an 'idealist' reading of scale, consciously chosing for "a semiotic approach" (Carr and Lempert 2016: 8), i.e. scaling that can be researched cross-culturally and cross-linguistically as a "scaling-as-sign activity" (2016: 10). Material affordances are not ignored but ultimately subsumed under 'meaning' following the conviction that "anything can be made big, brought near or perched atop a hierarchy" (ibid.). Carr and Lempert conclude that it is "not only that many aspects of social life can be and are scaled (space, time, politics, publics and interactions of all types); it is also that people employ different *senses of scale* when they engage in scalar practice" (Carr and Lempert 2016: 12). Arguably this raises an "anything goes" expectation with regard to scale and scaling. If the creative imaginary power of humans to think across scales in so many diverse ways is highlighted and if the main purpose was to document the scalar practices of others, why should scale constitute a problem for research beyond the documentation of semiotic complexity?

By and large anthropologists and archaeologists have sought to 'have the cake and eat it', i.e. continue using methods of small scale and individual sites while at the same time maintaining a claim on being able to develop theories of large scale. Archaeologists combined in-depth digging at single sites with modelling across sites and regions. Anthropologists insisted that they did not study villages but *in* villages, therefore they were dealing with "small places, large issues" (Eriksen 1995). How to get from small places to big issues, or from individual sites to general models for that matter, remained undertheorized and systematically underrated as a problem. Issues of sampling and statistical generalization were largely relegated to other disciplines, e.g. sociology, and questions of universals and their global spread remained a limited specialization within anthropology (see Antweiler 2007).

For the larger part anthropology was content to zoom in on the small scale when talking about the large scale: Single villages even in large societies such as Japan (Norbeck 1965), England (Ahmed and Mynors 1994) or Germany (Norman 1997) were assumed to be able to stand not only for the region or nation at large but also for a larger problematic such as industrialization or modernization. This way single cases came to stand for larger types. For example, 'the Nuer' came to stand for segmentary societies, 'the Hadza' came to stand for hunter-gatherers in marginal environments, 'the Yanomamö' came to stand for violent societies and so forth. The problematics have changed but the underlying strategies remained very similar. Often this involved not only a spatial upscaling or an upscaling from case to type but also a temporal scaling as cases of hunter-gatherers came to represent 'the stone age', just as cases of villages in industrial societies came to represent the rising era of globalization, the post-war period of industrialization, the transitions enforced by climate change in the Anthropocene and so forth. The underlying processes of scaling took place either invisibly or it was insufficiently reflected upon as an analytical problem but it nevertheless had considerable effects. A small group of !Kung could come to stand for human reciprocity or human affluence 'writ large' (see Wiessner 1982) and conversely "*Homo politicus*" could be researched 'writ small' by studying a group of Swat Pathans (see Barth 1959).

This state of affairs has only fairly recently been altered by attempts to reconcile 'materialist' and 'idealist' understandings of scale. Here three 'diagnostic' publications can help to illustrate how scale is currently being discussed – not only as a kaleidoscopic array of multiple worlds but as an issue that demands attention so that it provides analytical gain. Interestingly, all three publications deal with hunter-gatherers, people who had previously been seen as cases that could unproblematically be placed at the 'small-scale' end of the spectrum. The first two publications were extensively discussed during the workshop that gave rise to the current volume.

In the first relevant publication (Bird-David 2017a) Nurit Bird-David compared work with various "tiny" groups of foragers across continents which led her to diagnose anthropology with "scale-blindness". She complained that even though most ethnographers noted the smallness of the groups they were dealing with, they did not see the far-reaching implications that this had for the analysis but also for the people concerned. Although monographs and introductions to the discipline would note the small-scale of groups, they would still indulge in repeating and generalizing results from celebrated case studies independently of the fact that these groups were in fact tiny. The problem, it appeared, was bigger than simply a matter of statistical representativeness and the smallness of samples. Even though many of these groups would count no more than a few hundred people, it was taken for granted that upscaling was possible, and that it could be done in a number of directions. Bird-David problematized this seamless scaling up and down by arguing that the 'small scale' Nayaka that she worked with were very different and very unlike the same number of people in any larger society exactly *because* they were so few in number. Scale matters, she argues, because researchers living in large societies tend to assume that the social roles, rules and patterns they discover could also be identified in small-scale settings or may even allow us to see social entities and relations in their prototypical form. However, Bird-David states, it is not only that social roles differ across scales but it is questionable whether it makes sense to speak of 'social roles', 'persons' and 'individuals' at all under a certain threshold of smallness. Instead the "pluri-related" "pluri-present" and "pluri-connected" few that we find in these contexts, she maintains, are very unlike the "infinite few" in smaller subdivisions of large-scale societies (Bird-David 2017a: 215). The latter, she argues, can safely be multiplied as they are seen as "many beings of the same kind" (ibid.), similar to one another and deriving this similarity from being parts of a larger whole. For instance, a group of citizens in a nation state would receive their citizenship *qua* being singular instances of multiple members of that state. By contrast, in hunter-gatherer settings, Bird-David maintains, kin live with each other without being like each other (2017a: 219). They are 'pluripresent' (encounter one another in many intersecting relations) and at the same time they are diverse to the degree of being unique. This contrasts with large-scale societies that do allow members to be dispersed to the degree that they are unlikely to ever encounter everyone in the group while insisting on some sameness of members, having or assuming the same origin, same nationality, ethnicity or some other standardized shared feature (see 2017a: 217). What is at stake here is not only a distortion of life in small groups from the perspective of "modernity's large-scale horizons" (2017a: 215 see also 2017a: 210) but more generally a tendency to overlook what happens when we are scaling up or down in numbers.

Scaling not only refers to numbers and group size. Related but also distinct is the notion of scaling in extension and density. The question whether 'scale', understood as 'size', implies and connotes a whole series of related notions such as 'complexity', 'density', 'intensity', 'heterogeneity' and so forth had already been discussed by contributors to the Barth volume, e.g. by Berreman (1978). Berreman had noticed that many of the bi-polar ideal types such as the distinction between *Gemeinschaft* and *Gesellschaft*, folk versus urban, simple versus complex, personal versus impersonal and so forth, which have played a large role in the history of anthropology and social thought more generally, rely on implicit distinctions of scale. Many of these bi-polar distinctions, he argued, turn out to be distinctions of degree rather than of kind (1978: 70) but that this would still leave some "residual [...] analytical utility" to use scale in order to arrive at a typology of societies based on scale (1978: 50). This tendency to use measures of scale to place societies into a typology (of any sorts) has since been challenged on several accounts as exemplified by Bird et al. (2019), the second diagnostic text to which many contributors in this volume refer to. Douglas and Rebecca Bird, together with Brian Codding and David Zeanah (2019) combined ethnography with behavioural ecology as they revisited their long-term research with Martu Aborigines in Western Australia, another 'typical' case of small-scale hunter-gatherers (see Tonkinson 1991). They emphasize that although the group of Martu who would happen to be at the same place at any particular point in time would be rather small, these small residential groups were all connected through high and extensive mobility but also in terms of long-distance trading routes, ritual exchanges and marriages. Looking at sample numbers of average foraging groups (1-18 individuals, average 8.2, see Bird et al. 2019: 102), at average hearth groups (2-12 individuals, average 5.7) or average residential groups (41-127, see Bird et al. 2019: 101) is misleading, they claim. The authors conclude that despite residential and foraging groups being small "there is little evidence that these groups are drawn from small communities nested within small-scale societies" (2019: 96). Rather, they are dealing with "larger than expected local groups [...] maintained in expansive social networks of relational wealth involving interactions of hundreds of non-genetically related individuals" (2019: 106). There is a whole set of objections here against the earlier stance towards scale that Berreman and other contributors to the Barth volume, despite their uneasiness, were still ready to accept. Martu – and other hunter-gatherers like the Hadza and Ju/'hoansi referred to in this context – may live in small residential groups but at the same time they are also part of "complex and comprise large-scale networks" (Bird et al. 2019: 105). In other words, as social beings they are dealing with a number of relationships of quite different scale at any one point in time. Moreover, the residential groups are not of the static and tightly-knit type of close genetically-related kin that were often imagined. Instead, they were fluid and permeable so that the features previously associated with 'small-scale' do not necessarily apply, making such a typology rather useless. There is also "no discrete hierarchical scaling of three or more layers", the authors claim (Bird et al. 2019: 98), i.e. the groups or networks of different sizes do not exist as fixed and discrete layers of scale between which a society would 'switch'. Instead, the authors maintain, we are dealing with an "unbounded society" (Bird et al. 2019: 102). The actual groups of co-residences or co-workers cannot be predicted on the basis of any "well-defined community" (Bird et al. 2019: 94) – at any scale – from which they were considered to be drawn. The evidence the authors summon for these conclusions relies on both, the environmental usage patterns and the religious and ritual practices of the Martu, since both seem to defy a typology based on scale.

Bird et al. dismantle earlier assumptions that scale could be used to predict the shape of a society based on their mode of subsistence and that scale was an independent variable that could be applied to measure and predict their social relationships or modes of cultural adaptation to specific environments. Ironically, however, that does not lead to the conclusion that scale does not matter, but rather that modes of scaling are part and parcel of the social practices in various domains of life, including the social organization of co-residency and collaborative work, strategies of making economic use of resources in the environment and of living in a ritual and spiritual world of human and non-human agents. All this points to scaling as a practice: No matter whether it is population size, spatial expansion, density or some other dimension of scale that takes centre stage, it is in the nature of scale that has changed in the research process: It is no longer seen as providing a 'quick fix' to characterize or typify a society. Rather, the practices of scaling are now themselves subject to research. Scaling is not only routinely carried out by researchers but also by the researched. The third landmark publication that highlights this dimension of scale is Graeber and Wengrow's "Dawn of Everything" (2021) in which they show how scaling practices have been integral for social thought since the enlightenment but in which they also break with many assumptions about scale in the history of scholarly social research. Empirically, they suggest, throughout world history "most people live their lives on an ever-smaller scale as populations get larger" (2021: 141) which may be counter-intuitive against the background of current migration and globalization. In other words, there is an inverse dynamic here between regional and global networks growing in scale while the personal scales of movement are for the majority actually decreasing, at least spatially. In the emerging urban centres there is simply no need for the majority of people to travel as far as the Martu would need to do in order to satisfy everyday needs such as exchanging items or ideas (whether ritual or economical) or in order to find a spouse. In other words it seems that the 'overall' social complexity could be scaled up while the average spatial scale of individual moves is scaled down. The other main proposition by Graeber and Wengrow is that "our remote forager ancestors were much bolder experimenters in social form, breaking apart and reassembling their societies at different scales, often in radically different forms, with different value systems, from one time of year to the next" (2021: 140). In this perspective there is no unilinear development at all as people across time have been upscaling and downscaling their polities repeatedly. Upscaling and downscaling may be influenced by environmental factors of various sorts but these are scalar options chosen by the humans who find themselves in constantly changing situations. In fact, this is what Graeber and Wengrow make out as the main capacity that humans have (and non-humans don't), namely that they can switch between scales, both in terms of frames of orientation and in real-life institutions. Humans have, they maintain, for a long-time not only shifted seasonally and periodically between more dense/complex scales and less dense/complex scales. It is the cultural awareness that social relations could be taken to another scale which remained with them across these transitions and which continues to inform their strategies. In Graeber and Wengrow's view processes such as the 'Neolithic' domestication of plants and animals become a complex arrangement of scales as people (especially women) were experimenting with cultivating plants at a small physical scale ("play farming") but across a very long timescale (2021: 270). These small-scale ecological alterations over time would eventually lead to (largely unintended) large-scale changes as non-human species were tinkered and experimented with and as they were taken from one ecological setting to another. Moreover, Graeber and Wengrow break with the long-held assumption that hunting and gathering was necessarily associated with small-scale. Quite to the contrary, they argue, the scale of human sociality increased even before domestication was practiced at a large-scale. This relates to scale in terms of numbers (being urban) but also in terms of densities (being sedentary) and in terms of hierarchy (being non-egalitarian). One could go as far as saying that increasing scale is not a matter of recent processes of globalization becauset hunter-gatherers were also in the past living at the same scale of complexity as everyone else mastering the complex dynamics of shifting between decentralized and more egalitarian forms of organization and centralized, more hierarchical forms of social organization (Graeber and Wengrow 2021: 314).

The argument that hunting and gathering was one strategy practiced alongside (and alternating with) other modes of subsistence and that the transitions were not simply a function of population growth had been aired before from within the field of hunter-gatherer studies (see Layton et al. 1991). However, Graeber and Wengrow not only advocate a change in analytical perspective but they also connect this to a political agenda and with a critique of strict versions of evolutionary behaviourism. Thereby they undermine the assumption that humans were "naturally" equipped to deal only with small-scale social relationships. They question the assumption that any increase in scale that we see in recent history (larger numbers, permanent settlements etc.) necessarily means that dominating structures had to be put into place (2021: 310) as if, for instance, participatory democratic forms of organization were impossible at a larger scale (2021: 573). The evidence they summon for this position is the observation that many hunter-gatherers live at two scales simultaneously, either in terms of seasonality (see Wengrow and Graeber 2015, going back to seminal ideas by Marcel Mauss) or in terms of their flexible mindset (see Graeber and Wengrow [2021: 314] where they refer to the Bird et al.). What they seek to add to the materiality of changes in scale (population growth) is that scale is a dimension of the mind (2021: 314). Humans imagine cities before and independently of constructing them. This adds yet another dimension to our research on scale. Scale not only has a number of dimensions (quantity, size, density, complexity etc.) but it is also a standard and gauge that is applied as a part of social practice. Moreover, it is not only researchers who apply these standards, for instance when they label a society as being 'small-scale', but humans do so continuously in the process of living their lives. As hunter-gatherers, or as industrial workers for that matter, they set scales which in turn has implications for their behaviour and for the landscapes and polities that emerge as a consequence. For most contemporary hunter-gatherer groups, for instance, it would be true to say that they have upscaled hunting and gathering from subsistence pursuits to an integrated cultural way of life that goes beyond subsistence activities and which is better understood relationally in terms of a hunter-gatherer situation (see Widlok 2016). This brings us back to the issue raised by Nurit Bird-David: It is not only that western scientists brought up in modern societies are in the danger of mis-representing other people living their lives at different places or different times. Bird-David's Nayaka interlocutors, too, are applying their 'kin scale of plurirelational beings' as a frame for their own actions and for orienting themselves. And if Graeber and Wengrow's argument holds, Nayaka or any humans living together are capable of doing what architects do when they provide models of what is to be built at a variety of scales (e.g. 1:2000, 1:200, 1:20). We all can scale our relations with one another and with non-humans in a similar way. We not only live in different scales, we can also *apply* different scales. We have every reason to believe that this is true for every group of humans that features in the contributions of this volume, no matter where or when they lived. It is not only that 'they' (the researched) scale as much as 'we' (the researchers) scale. Rather, 'we' have learned our lessons of scaling from a long history of scaling that we encounter whenever we turn to a group of 'they' in the archaeological record and in anthropological case studies. Scaling is a complex practice but getting it right is highly relevant not only for understanding ways of living at other times and in other places, but ultimately for exploring the potentials of living our human lives in an environment of multiple scales.

In the sense outlined above it becomes clear that the case studies collected in this volume are really all variants of one single and ongoing case, that of human scaling practice that characterizes the human condition. We have tried to emulate this realization in the way that we present the individual contributions to this volume. The contributions need not be read in any particular order since there are cross-references throughout. The sequence of articles in this book is more or less that of the meeting that we had in 2020 and which was sponsored by the Collaborative Research Center 806 "Our Way to Europe" at the University of Cologne. We acknowledge the support of the *Deutsche Forschungsgemeinschaft* that funded the CRC, enabled our meeting and subsidized this publication. We are also grateful to the contributors who came together under the adverse conditions of the pandemic and who were prepared to develop their ideas, discuss each other's contributions and bring this project forward despite the difficulties. We also appreciate the input that Robert Layton (Durham) and Andreas Womelsdorf (Vienna and Heidelberg) provided as discussants during our meeting. Souhayb Zaryah provided invaluable technical support in organizing the online workshop and in preparing the manuscript. When preparing this volume, we took a stance that underlines the distinctive contribution of each author. Hence we have not harmonized the 'Englishes' that are being used and we have been careful not to flatten out the diverging (sub)disciplinary perspectives. During the conference all contributors emphasized how rarely scholars of such different backgrounds actually come together to exchange views in a dialogical rather than a confrontational matter. It was felt by the participants that adding some individual comments to the chapters across the volume would be an appropriate way to capture the constructive atmosphere that the meeting had. Each contribution in this volume is therefore followed by one or two comments by fellow authors. Thereby we try to counteract the growing unease that the subfields of anthropology, and the dominant theoretical currents within the field at large, are drifting apart at a speed, and – dare we say – at a scale, that is detrimental to solving large research questions such as those that are being dealt with in this volume. At a meta-level we therefore hope that we have shown that as specialized scholars we are still in the position to scale our findings and ideas in a way that not only responds to the specific research approaches that we are particularly committed to but that we are also able to collaboratively scale-up when facing the larger challenges of making sense of human life.

References

Ahmed, Akbar and James Mynors 1994. Fowlmere, Roundheads, Rambo and Rivalry in an English Village Today. *Anthropology Today* 10/5: 3-8.

Antweiler, Christoph 2007. *Was ist den Menschen gemeinsam? Über Kultur und Kulturen*. Darmstadt: Wissenschaftliche Buchgesellschaft.

Barth, Fredrik 1959. *Political Leadership among Swat Pathans*. London: Athlone.

Barth, Fredrik (ed.) 1978. *Scale and social organization*. Oslo: Universitetsforl.

Berreman, Gerald 1978. "Scale and social relations". In: F. Barth (ed.) *Scale and social organization*. Oslo: Universitetsforl., pp. 41-77.

Bird-David, Nurit 2017a. Before Nation: Scale-Blind Anthropology and Foragers' Worlds of Relatives. *Current Anthropology* 58/2: 209-226.

Bird-David, Nurit 2017b. *Us, Relatives: Scaling and Plural Life in a Forager World*. Oakland: University of California Press.

Bird, Douglas W., Rebecca Bliege Bird, Brian F. Codding, and David W. Zeanah 2019. Variability in the organization and size of hunter-gatherer groups: Foragers do not live in small-scale societies. *Journal of Human Evolution* 131: 96-108.

Brenner, Neil 2001. The limits to scale? Methodological reflections on scalar structuration. *Progress in Human Geography* 25/4: 591-614.

Carr, E Summerson, and Michael Lempert 2016. *Scale: Discourse and dimensions of social life*. California: University of California Press.

Graeber, David and David Wengrow 2021. *The Dawn of Everything. A New History of Humanity*. New York: Farrar, Straus and Giroux.

Howitt, Richard 1998. Scale as Relation: Musical Metaphors of Geographical Scale, *Area* 30/1: 49-58.

Layton, Robert, Robert Foley and Elizabeth Williams 1991. The transition between hunting and gathering and the specialized husbandry of resources. A socio-ecological approach. *Current Anthropology* 32/3: 255-263.

Marston, Sallie 2000. The social construction of scale. *Progress in Human Geography* 24/2: 219-242.

Marston, Sallie and Neil Smith 2001. States, scales and households: limits to scale thinking? A response to Brenner. *Progress in Human Geography* 25/4: 615-619.

Norbeck, Edward 1965. *Changing Japan*. N.Y.: Holt, Rinehart & Winston.

Norman, Karin 1997. *Kindererziehung in einem deutschen Dorf. Erfahrungen einer schwedischen Ethnologin*. Frankfurt/M.: Campus.

Tonkinson, Robert 1991. *The Mardu Aborigines. Living The Dream in Australia's Desert.* 2nd ed. Forth Worth: Holt, Rinehart & Winston.

Wengrow, David and David Graeber 2015. Farewell to the childhood of man: ritual, seasonality, and the origins of inequality. *Journal of the Royal Anthropological Institute* 21/3: 597-619.

Widlok, Thomas 2016. Hunter-gatherer situations. *Hunter gatherer research* 2/2: 127-143.

Wiessner, Polly 1982. "Risk, reciprocity and social influences on !Kung San economics". In: E. Leacock and R. Lee (eds.), *Politics and history in band societies*. Cambridge: Cambridge University Press, pp. 61-84.

Xiang, Biao 2013. Multi-scalar ethnography: An approach for critical engagement with migration and social change. *Ethnography* 14/3: 282-299.

How do we scale hunter-gatherers' social networks?
Towards bridging interdisciplinary gaps

Nurit Bird-David

For the greater part of the 20th century, hunter-gatherer societies were regarded as "small-scale societies" by modern sociology (see Barth ed. 1978 for overview). The broad distinction between "small-scale" and "large-scale" was increasingly losing its overall analytical grip in the late 20th century, also with regard to the classification of hunter-gatherers in these terms. Anthropologists who approached hunter-gatherers from ecological-evolutionary perspectives continue to address group-size as a key explanatory issue (see Kelly 1995 for overview), and some even associated it with the evolution of the human social mind (Dunbar 1993). By contrast, anthropologists who approached hunter-gatherers from socio-cultural perspectives have for the most ignored group – and population – sizes, regarding demographic figures as marginal to understanding hunter-gatherer societies, cultures and worlds (see Bird-David 2017 b, 2018, 2019 for overview). I, too, have been one of those latter ethnographers until I recently changed my approach to consider manifold and complex aspects of "scale" as at once concept, phenomenon, approach and much more (see Carr and Lempert 2016 for overview of the concept of scale).

What made me reconsider my approach was writing a book on the forest-forager Nayaka of South India, with whom I have been working since 1978 (and my students Daniel Naveh and Noa Lavi respectively since 2003 and 2010). My original plan was to write an ethnographic monograph, after having published many articles. I wanted to write a monograph from the standpoint of my long-term perspective of close to four decades of work with Nayaka. But the project became more complicated than I initially envisaged. The more I wrote, the more it dawned on me how any ethnographic description and analysis of Nayaka culture (and that of hunter-gatherers more generally) fails to evoke their experiences lest we pay attention to scalar aspects of their lifeways. That is, we need to pay attention to how we, and how they, scale and imagine societies and lifeworlds, and the affordances and limits entailed in these scales. Consequently, I aligned my project with turn-of-the-21st century work in the social sciences concerned with "scale" as a modern analytical and discursive concept (what has since been tagged "the scalar turn", see in Carr and Lempert eds. 2016). And I integrated scalar issues into my book (Bird-David 2017a) and into follow-up articles (Bird-David 2017b, 2018, 2019). I started to ask myself who scales hunter-gatherer societies, by what criteria and for what purpose, and whether and how they scale themselves. I attended to issues of scales starting from the Nayaka order of magnitude, which comprises a few dozen people for residential groups, a few hundred for local communities, and rather unreliable outsiders' estimates of a few hundred to a few thousand for the entire population. These figures are comparable with those for many (not all) other hunter-gatherers. I went further from the numbers and explored how – as far as the order of magnitude goes, in comparison with other societies, and especially modern Western societies – hunter-gatherers' small order of magnitude influences their lifeways and lifeworlds. In my recent work, I revisited a range of topics in hunter-gatherer studies from this scale-sensitive perspective, e.g., kinship, marriage, gender and child rearing, relations with nonhumans, outsiders and the state, animistic and relational ontologies, and more. I showed how relevant is "smallness" of hunter-gatherers' social aggregates, and the "nearness" of the horizons of their worlds, are to understanding their lifeways and culturally-created worlds.

An apparently opposing perspective on hunter-gatherer scale has meanwhile been offered from an ecological-evolutionary standpoint by Douglas Bird and associates (Bird et al. 2019). These ethnographers drew on their work with Australian Aboriginal Martu people since 2000, as well as comparative work among Hadza and others. They took an evolutionary-ecological perspective, and argued that "foragers do not live in small-scale societies" (the article's subtitle). To the contrary, they argued, foragers live in "large-scale social networks" (p. 69). Based on this revision of hunter-gatherers' scale, the authors proposed – contra Dunbar's famous argument (1995) – that human cognition coevolved with large-scale social networks, with socio-ecological interactions and relational wealth.

On the face of things, we have conflicting assessments of the scale of hunter-gatherers social aggregates – in short referred to below as "sociocultural" vs. "ecological-evolutionary" perspectives and "very-small-scale" vs. "large-scale" positions. Up until today, these arguments have not been thoroughly debated, the main reason being a deepening disciplinary split within hunter-gatherer studies. Hunter-gatherer studies was constituted in the 1960s as a comparative and interdisciplinary project (Lee and DeVore eds. 1968), not least because ethnographically-observable hunter-gatherer cases are scarce, and their comparative study is crucial to understanding their recurring patterns and their variations, and in turn critical for speculating about our past and about human evolution. Ironically, the more sub-disciplinary traditions of socio-cultural versus ecological-evolutionary research on hunter-gatherer advanced since the 1960s, the more these traditions drifted apart beyond each other's range of comprehension – paradoxically precluding a comparison of cases which is so essential to the overall project. However, debating these two apparently contradictory assessments of hunter-gatherers being "small-scale" versus "large-scale" is important because this is not simply a matter of group size. Rather, what is at stake are wrong and misleading practices of scaling when theorizing on modern hunter-gatherer worlds (in the first case) and on human evolution (in the second case). We should therefore make an effort and debate these two arguments together, and in the process we may also hope to contribute a little to narrowing the disciplinary gap between the socio-cultural and ecological-evolutionary traditions in hunter-gatherer studies. This is precisely one of the main objectives of this collective volume, and in my contribution I want to take up this challenge.

In this chapter, I will ask if the socio-cultural and ecologically-evolutionary positions (foragers live in "very-small" vs. "large" scale social worlds) are really as contradictory as they initially seem to be. I will argue that these arguments accord with each other far more than their rhetoric and argumentative style may lead us to believe. The demographic figures upon which these arguments are based, I will show, do not necessarily contradict one another. Only their rhetorical package as "small" and "large" suggests contradictory arguments. Classifying the same demographic order of magnitude as "small-scale" (in the first instance) and as "large-scale" (in the second instance) may be the result of the different readerships that are being addressed: In the first case, I address socio-cultural ethnographers who compare hunter-gatherers with modern western societies while ignoring hunter-gatherers' comparatively tiny scale. In the second case Bird et al. (2019) address ecological-evolutionary students who still adhere to the worn-out "small-scale society" stereotype. However, as I hope to show in this chapter, these arguments do not necessarily pull us in opposite directions.

I start this chapter by opening up the basic terms we use, including population size, scale and more generally the quantification of hunter-gatherers' social forms. My aim is to explain why socio-cultural ethnographers avoid quantification and yet I shall urge us not to do so for a better understanding of hunter-gatherer worlds and for the benefit of interdisciplinary hunter-gatherer studies. Next, I turn to the "large-scale" claim of ecological-evolutionary ethnographers, and show that it rests not so much on their actual quantitative figures but on how these figures are verbalized, interpreted and assessed through figures of speech. As I shall show this includes the construction of strawmen, of binaries and ethnocentric definitions. Showing that these two positions factually agree more than disagree with each other, and that the seeming conflict between them reflects on our insufficiently refined analytical terms, I propose an analytical refinement in the conclusion, consisting of three steps. First, at its simplest, a shift from binary reading of small and large scales to their relational reading as gradients on a scalar continuum. Second, analysis predicated on hunter-gatherers' social networks, rather than "societies". Third, a discussion of the intensity and density of hunter-gatherers' social networks', rather than just of their scale, with attention to their modes of sociality *and* subsistence, and the spatial correlates.

Mind the gap: Quantitative figures

Scale is a complex word. Dictionaries alone list multiple different meanings. In the social sciences, this term has been used for close to a century as a key analytical concept. The concept of scale includes the size of groups and populations but goes much further and associates the sizes with different social systems (polities, economics, cultures, etc.) within a grand binary modern distinction between "small-scale societies" and "large-scale societies" (see Barth 1978 for extensive discussion of this distinction). In this section, I want to press home that scale is a complex concept with multiple and changing meanings. Even in its apparent simplest sense as size of local groups and of societal total populations, questions arise as to who should be counted, by what criterion, when, by whom, and for what? Population surveys are entangled with identity politics and build on epistemological and ontological assumptions that are far from trivial and universal. All the same, socio-cultural ethnographers cannot simply ignore figures.

Even if we refer here to scale in its basic sense, group size, the scale of hunter-gatherers' social aggregates is a complex matter. A child can count several dozen members, the order of magnitude claimed to be typical for hunter-gatherer bands. But is it that simple? Sure, if I had the opportunity to time-travel with my grandson to the Nayaka I studied, he could easily count all those who lived in the hamlet I lived in: 28 men, women and children, 69 in the five hamlets who kept visiting each other and stayed in close contact. The problem would still be as to who should be counted? Everyone sleeping at the hamlet on a selected night, or present in the hamlet at a certain moment? Children of mixed marriages (e.g., of a Nayaka woman and a Muslim man) living in the hamlet along with those with parents who are both Nayaka? What about someone who left the hamlet a day before we do our count to visit and stay with relatives in another hamlet, often for an indeterminate duration of time? Or, the family who only arrived two weeks ago and are still staying in the hamlet? Other things complicate the matter: Should the hamlet's dogs be counted? Or young wild animals adopted and taken care of as a sort of children? And, allowing for polemically unsettling our assumptions and biases, should a couple who always stay together, or a mother and her just-born baby, be counted as two or, maybe, only as one social entity? The issue that these questions raise is not simply technical in nature. Technically, these questions can be resolved by the researcher's arbitrary decisions appropriate to his or her particular case. Rather, these questions begin to raise epistemological and ontological problems concerning the scaling of hunter-gatherer groups. As we move beyond the local group to their regional aggregates and the entire population, the basis of counting the hunter-gatherer population becomes more intricate as I want to explain in more detail below.

The composition of local groups is in constant flux, and groups keep moving from place to place. Even if this is solved by armies of surveyors with sufficient time at their disposal, the question of what criteria are to be employed remains. Who decides on the criteria, who reaches their scattered settlements in the wild in order to count them? Most estimates that appear in the hunter-gatherer literature have been produced by outsiders, commonly colonial and state administrative staff and sometimes missionaries and explorers. Even if we assume that they did their job well, itself a daring assumption, their estimates are entangled with identity politics and economics. The estimates are based on politically – and economically – motivated practices of naming and classifying peoples by their ethnicity, religion, language etc. In the case of many indigenous populations, even their ethnonyms are chosen by outsiders and often change from time to time and between those who name them (Bird-David 2017b).

Population surveys aim to assist governing large-scale systems. The colonial India's population survey, for instance, which began in the early 19th century, is second to none for its massive scale and intricate classification. Continuing into independent India's national surveys, the 200 years old series of surveys show how inconsistent the naming and the enumeration of small so-called "tribal communities" like the forager Nayaka has been (Bird-David 2017a). Moreover, the very idea of the "population at large" assumes a notion of society as a category comprising of individuals who – irrespective of whether they know or engage with one another – are "alike" with regard to this or another criterion. As an "imagined community" (Anderson 1991[1983]), societies are aggregated in the mind, or be it on paper or with the help of a computer, as a clear-cut group and category. On the ground, people with diverse and complicated biographies and histories have to be "pushed" into this or that group category so that they can be counted. Against this background one begins to understand why social-culturally trained ethnographers of hunter-gatherers have, for some decades now, turned a blind eye to the scalar framework of hunter-gatherers' worlds. At best they mentioned demographic figures only in a line or two when introducing the people with whom they conduct research in their case studies. The sources were not very reliable. Unfortunately, they have not dealt with analytical implications of largely disregarding the problem.

At the same time hunter-gatherer demographic figures are of great importance for ecological and evolutionary perspectives. This is why much effort has been invested by scholars of these approaches to procure and collate hunter-gatherer demographic figures. Several scholars painstakingly collated figures from sources of all kinds, administrative and ethnographic, going back to 19th century sources and sifting through socio-cultural ethnographies. There is a three-page long table produced by Robert Kelly (1995: 206-8) and a seven pages long table produced by Lewis Binford (2001: 245-251) which are exemplary for this effort. Again and again these have since been cited in ecological-evolutionary work, and the more they are cited the more their authority is established. Unfortunately, their tenuous basis, including the fact that figures were taken over from colonial sources, socio-cultural ethnographies, etc., have been "forgotten" in the process. Talking across the disciplinary gap in hunter-gatherer studies, could greatly benefit by no longer ignoring hunter-gatherers' demographic figures and by addressing the issues surrounding the epistemological and historical basis of these figures.

Mind the gap: Figures of speech

Having discussed the quantification of figures across the disciplinary spectrum, I now want to underline that caution needs to be taken when expressing these figures, with equal attention given to the figures of speech that are used. I turn to Bird et al. (2019) for illustrating what I mean. The wealth and quality of the quantitative figures that they provide, together with their radical claim that "foragers do not live in small-scale societies" which is based on these figures, lends itself to illustrate my point.

The authors draw on long-term, extensive ethnographic fieldwork with the Australian Desert Martu people. Their research teams have been attentive to quantification and they systematically produced quantitative data in a way that socio-culturally oriented ethnographers, who commonly work alone, do not, (nor can) usually do. Their claim discussed here is based on data collected between 2000-2010, compiling data on the composition of foraging groups, on the amount of time invested in foraging and the yields produced. It also includes data collected for a period of 8-weeks during 2010 on residential group fluctuation in a particular locality. Additionally, the authors turned to "basic census data" collected by Welfare patrol officers, who contacted isolated Martu groups in the 1960s, and they interviewed living Martu members of those 1960s groups. This is a commendable data basis by all accounts, but I want to show, that it supports (and not contradicts) the "smallness" of hunter-gatherers' worlds. The actual data, I argue, is occluded by the authors' choice of rhetorical and argumentative figures of speech.

The first rhetorical obstacle is that of setting up a strawman. Based on their data, the authors critically address the assumption that "groups of co-residents are nested within small communities that are, in turn, nested within small-scale societies" (p. 96). Some scholars outside hunter-gatherer studies may still be subscribing to this model but it is important to emphasize that students of hunter-gatherers have long emphasized the "flux" and "fluidity" of hunter-gatherers' groups, at least since the 1960s. The fluidity of local groups was, in fact, celebrated as one of the major conclusions reached in the 1966 conference "Man, the Hunter," the cross-disciplinary conference that started modern hunter-gatherer studies. Moreover, at least since then no ethnographer of hunter-gatherers has claimed what Bird et al. critique in their article, by means of their good data, namely that "well-known hunter-gatherers do not live in hierarchically organized, small-scale societies" (p.97). To my mind this is a wasteful use of invaluable data, because it is used here to knock-down a strawman and it diverts attention away from the really critical issues.

The second rhetorical obstacle is their usage of "small" and "large" as binary terms in a way that in fact misrepresents the data. What are the actual figures that are provided in the article on Martu and other hunter-gatherer group size? Bird et al. carefully distinguish between four group-levels: a) "hearth groups", small family groups spatially spread around the settlement's center; b) "residential groups", the hearth-groups living around the same center; c) "foraging groups", people who day-forage together; and d) "large-residential groups" (called *tjapal* by Martu), gatherings taking place now and then during the year for ritual and social "business." For each group level they provide data on group size: 3-10 individuals for hearth groups, 41-127 for residential groups, 1-18 for foraging groups, and "upwards of hundreds of people" for the "large residential group" (pp. 101-103).

Being larger than the other smaller Martu groupings, it is the largest grouping with its "upwards of hundreds of people" which is tagged as "large residential group" and it is given particular comparative attention. The authors show similar group-sizes among other hunter-gatherers. For example, they cite studies of Hadza and Aché showing that adults typically interact with "hundreds of other adults during their life time" and are likely to observe "over 300 different men making tools over the course of their lives" (Hill et al. 2014, cited by Bird et al. 2019: 98). They cite Blurton-Jones who, based on a 15 years long survey of the Hadza, wrote that it is "completely wrong" to think of them as tiny bands averaging 21 people (ranging from 20 to 100) since each person recorded had co-lived with an "astonishing average of 69 different people" in the camps he moved between during 15 years (Blurton-Jones 2016, cited by Bird et al. 2019: 98). Figures of a similar order of magnitude can be added here, cited by David Wengrow and David Graeber (2015) for their own separate theoretical ends that need not concern us here beyond stating that these authors argue that hunter-gatherers alternated between small egalitarian organizations and large hierarchical political organizations. Wengrow and Graeber draw on 19th and early 20th century literature on North American hunter-gatherers. They cite, for example, Mauss and Beuchat seminal *Seasonal Variations of the Eskimo: A Study in Social Morphology* (1979 [1904-5]), which examines annual shifts between summer and winter settlements. In the summer, individual families lived in tents, dispersed and scattered over an immense area.

In the winter, the families congregated in "large" concentrated settlements of multi-family and communal houses to perform collective ceremonies. Mauss and Beuchat painstakingly researched contemporary and earlier surveys and concluded that the winter settlements, or what are elsewhere called the "large" groupings, consisted of eight to fifteen houses comprising 200 to 400 members.

Now, "upwards of hundreds" is surely larger than hunter-gatherers' local group size of several dozen people. A "few hundred" is surely larger than the "magic numbers" of local groups recognized in "Man, the Hunter" and of what is widely endorsed today, namely "25-50" men, women and children living in the same camp. But from here a slippery binary verbal slope leads to arguing that foragers do not live in "small" groups and, then, that they live in large assemblies. It is a slippage from "not-small" that is turned into "large". That hunter-gatherers often do not live in groups as small as those that ethnographers focused on for too long, does not automatically mean that they live in large-groups. Only under the tyranny of a binary split between "small-scale" and "large-scale" society does "not small" automatically turn into "large." And "hundreds" of hunter-gatherers suddenly figure as "large-scale," along with large-scale modern societies of hundreds and thousands of millions in the same category. The valuable data provided by Bird et al. meanwhile is misused as a basis for arguing, as the authors do, that foragers do not live in "small-scale" societies but, instead, in "large-scale" social networks of interaction (I return below to the insightful shift from "society" to "social network"). To the contrary, I want to argue that their data in fact strongly presses home the comparative "smallness" of hunter-gatherers' social formations, even at their largest reach "upwards of hundreds." Their data actually supports my socio-cultural argument that ignoring hunter-gatherers' scale, even in its simple sense of demographic order of magnitude, along with what it limits and affords, obstructs our understanding of foragers' lived-experiences and worlds.

The last rhetorical obstacle on the way of bridging the gap between the different scholarly traditions in hunter-gatherer studies concerns the question of kinship relations. Are members of a hunter-gatherer group mostly kin or rather non-kin? Along with other ecological-evolutionary oriented students, Bird et al. (2019: 96) argue that "most mobile hunter-gatherers live in groups dominated by links between non-relatives." Their argument appears to radically turn the earlier consensus on its head according to which kinship is the basis of hunter-gatherer bands. The kinship basis of the hunter-gatherer band was assumed from the 1930 to the 1970s, kinship then constituted an important topic, and arguments revolved over which type of kinship relation characterizes the composition of the band. In the wake of "Man the Hunter", the interest then shifted to issues of the "hunter-gatherer mode of subsistence" (see Bird-David 1995 for more details). Thereafter, little by little, a few socio-cultural ethnographers of hunter-gatherers returned to kinship as a cultural phenomenon (see Bird-David 2017a). Against this background, we can revisit the polemic ecological-evolutionary argument that it is mostly non-relatives that comprise a hunter-gatherer group.

The "non-kin" argument resonates with that of the hunter-gatherer "large-scale" social formation, and likewise it is trapped within binary opposites. Bird et al. (2019), may serve as an example of other ecological-evolutionary statements on this issue. They, limit what they count as kinship connections to a "coefficient of relatedness greater than 0.06" (2019: 103). This scientific index limits those considered as kin just up to second cousins. By this definition, anybody else is non-kin, notably including relatives through marriage. This definition clearly departs from hunter-gatherers' own sense of kinship – but also from what is commonly regarded as kin in daily life by many Western people! While we may, arguably, put aside decades of contact with the state, including leaving and returning to desert settlements when examining foraging parameters, the same could hardly be done when examining genetic kinship connections among Martu people living in Government settlements in the 2000s. For Martu people, we learn from the authors, as for many other hunter-gatherers, affinal ties are important kinship ties. Marriage ties connected many members of Martu groups in the 2000s (Bird et al. 2019: 102). In the 1960's, we learn from their ethnographer Robert Tonkinson (2004), kinship was central, and it had even been the idiom through which Martu established relations with outsiders. All of these manifold kinship relations are not counted as kin by the authors. There is no sense of kinship being a gradient between close and more distant but there is instead a categorical cut between kin and non-kin.

All in all, the figures of speech outlined above occlude what the quantitative figures show in the article by Bird et al. (2019) and that I have referred to here as an example. If we remove the binary scaffoldings from the scalar augments there is hardly a basis for concluding that "foragers do not live in small-scale societies," that they "live in large-scale social networks," and their members are largely "non-relatives." Having said that, the article, at the same time, convincingly suggests that we should shift from "society" to "social network" as the overall theoretical construct. The polemic arguments on hunter-gatherers' scale may overshadow this proposal which is why I want to highlight its implications. For me, the move from "society" to "social networks" seems a promisingly productive way to approach hunter-gatherer sociality across the disciplinary gap in hunter-gatherer studies.

Concluding remarks: Towards refining our analysis of hunter-gatherers' social networks

So far I suggested that there is no real conflict between the argument derived from socio-cultural anthropology that smallness is analytically essential for understanding hunter-gatherer social worlds, and the argument derived from ecologically-evolutionary anthropology that hunter-gatherers "do not live in small-scale societies". The illusion of their conflict is created when socio-cultural anthropologists continue to doubt and underuse quantitative figures, and when polemic figures of speech in ecological-evolutionary anthropology cloud the data. Going beyond what are only seemingly discordant scalar claims, I suggest to move forwards by exploring hunter-gatherers' social networks.

This middle position involves shifting the focus from the smallest to the largest hunter-gatherer group levels, yet admitting that even the largest level is still "small" as far as scale goes in comparison when including the entire diversity of human societies. It involves simultaneously shifting from "society" to "social network," a sociological concept traced back to the work of Simmel (1950[1908]) and later operationalized as a sophisticated conceptual and methodological package, and which was originated and developed largely within the context of studying large-scale modern society. The challenge for scholars of hunter-gatherers is to adapt concepts and tools of "social networks" to the hunter-gatherer small-scale world, rather than apply the range of given tools to the hunter-gatherer and misrepresenting them as "large-scale social networks." The suggested turn towards the notion of social networks suggested by ecological-evolutionary hunter-gatherer scholarship can be brought together with a recent turn in socio-cultural hunter-gatherer scholarship towards relations and relationality as keys to understanding the hunter-gatherer culture and world. In the ecological-evolutionary approach relevant work on hunter-gatherer social networks include Apicella et al. (2012), Hamilton et al. (2007), Migliano et al. (2017), and Whallon (2006). In the socio-cultural approach on relations and a relational perspective this includes Myers (1986), Bird-David (1999, 2017a), and Ingold (2000).

"Intensity" as a property of social networks is a case in point and may be a good direction for developing this train of thought in further interdisciplinary work. Low-density population is a condition of successful subsistence based on hunting and gathering natural resources. At the same time, hunter-gatherers' social relations depend on performing them, rather than just knowing them, in other words what counts here is connecting with others by being-with them rather than by mapping relations against a genealogical template. Social relations have to be constantly reproduced and reaffirmed in order to be recognized and counted by hunter-gatherers. This is partly why hunter-gatherers constantly visit each other, why they share food, space and in a sense their selves. And this is why their groups are fluid, why they move between aggregates far more than ecological/economic factors can explain, and why now and then they gather in large residential groups, although their subsistence needs are met better when they live in small groups. Their social groups exist through their members' intense interactions with each other.

The ingenuity of hunter-gatherer social organization, I suggest, is articulating low-density population and high-intensity interaction so as to subsist and exist as individuals and as a collective. Intensity is the solution to the hunter-gatherers' paradox: low-density population for maximizing subsisting on natural resources, and high-intensity interaction for keeping their social networks going. This leads to suggesting several points to think about and pursue in future research. Instead of calling them "local groups" or "residential groups," they are rather approached as social networks, too. The "local" and the "regional" social networks can then be discussed by comparing their intensity, in relation to their social and subsistence practices as and when productive, the gradient social networks can be compared by such social network key terms as multiplexity to describe multiple ties between members, and propinquity to emphasize its correlation with members' geographical closeness. The dynamic articulation of hunter-gatherers' gradient social networks from residential to wider social networks could provide a basis for including space/territory in the analysis and discussing long-term processes of spatial and population expansion. Going beyond hunter-gatherers' social organizations, instead of calling their networks either small or large, they are rather characterized by a specific density. What we have called hunter-gatherers' "small scale society" then can figure as social network with low density subsistence and high density sociality, and what we have called "large societies" would rather be high density subsistence with low density sociality. But in both instances understood as gradients that are subject to change by both external ecological conditions and internal socio-cultural reasons.

References

Anderson, Benedict. 1991. *Imagined communities: Reflections on the origin and spread of nationalism*. 2nd ed. New York: Verson.

Apicella, Coren L, Frank W Marlowe, James H Fowler, and Nicholas A Christakis. 2012. Social networks and cooperation in hunter-gatherers. *Nature* 481 (7382): 497-501.

Barth, Fredrik. (ed.) 1978. *Scale and social organization*. Oslo: Universitetsforlaget.

Binford, Lewis R. 2001. *Constructing frames of reference: An analytic method for archaeological theory building using ethnographic and environmental data sets*. Berkely: Univ of California Press.

Bird, Douglas W, Rebecca Bliege Bird, Brian F Codding, and David W Zeanah. 2019. Variability in the organization and size of hunter-gatherer groups: Foragers do not live in small-scale societies. *Journal of human evolution* 131: 96-108.

Bird-David, N. 1994. Sociality and immediacy: Or past and present conversations on bands. *Man* 29 (3): 583-603.

Bird-David, N. 1999. Animism revisited: Personhood environment and relational epistemology. *Current Anthropology* 40 (Supplement):S67-91.

Bird-David, N. 2017a. *Us, relatives: scaling and plural life in a forager world*. Oakland, CA: Univ of California Press.

Bird-David, N. 2017b. Before nation: Scale-blind anthropology and foragers' worlds of relatives. *Current Anthropology* 58 (2): 209-226.

Bird-David, Nurit. 2018. Size matters! The scalability of modern hunter-gatherer animism. *Quaternary International* 464: 305-314.

Bird-David, Nurit. 2019. Kinship and scale: On paradoxes in hunter-gatherer studies and how to overcome them. *Hunter Gatherer Research* 4 (2): 177-192.

Carr, E Summerson, and Michael Lempert. 2016. *Scale: Discourse and dimensions of social life*. Oakland, CA: Univ of California Press.

Dunbar, Robin. 1993. Coevolution of neocortical size, group size and language in humans. *Behavioral and Brain Sciences* 16: 681-735.

Hamilton, Marcus J, Bruce T Milne, Robert S Walker, Oskar Burger, and James H Brown. 2007. The complex structure of hunter–gatherer social networks. *Proceedings of the Royal Society B: Biological Sciences* 274 (1622): 2195-2203.

Ingold, Tim. 2000. *The perception of the environment: Essays in livelihood, dwelling and skill*. London and New York: Routledge.

Kelly, Robert L. 1995. *The foraging spectrum: Diversity in hunter-gatherers' lifeways*. Washington and London: Smithonian Institution Press.

Lee, Richard. B., and Irvine DeVore (eds.). 1968. *Man the hunter*. Chicago: Aldine.

Migliano, Andrea Bamberg, Abigail E Page, Jesus Gómez-Gardeñes, Gul Deniz Salali, Sylvain Viguier, Mark Dyble, James Thompson, Nikhill Chaudhary, Daniel Smith, and Janis Strods. 2017. Characterization of hunter-gatherer networks and implications for cumulative culture. *Nature Human Behaviour* 1 (2): 1-6.

Myers, F. 1986. *Pintupi country, Pintupi self: Sentiment, Place, and Politics among Western Desert Aborigines*. Washington: Smithonian Institution Press and Australian Institute of Aboriginal Studies.

Simmel, George. 1950. *The sociology of George Simmel*. Translated by K.H. Wolff. NY: Free Press.

Wengrow, David, and David Graeber. 2015. Farewell to the 'childhood of man': ritual, seasonality, and the origins of inequality. *Journal of the Royal Anthropological Institute* 21 (3): 597-619.

Whallon, Robert. 2006. Social networks and information: Non-"utilitarian" mobility among hunter-gatherers. *Journal of anthropological archaeology* 25 (2): 259-270

Comment by Charlotte Damm

In numerous recent publications Nurit Bird-David advocates strongly for the necessity to take the "small-ness" of many hunter-gatherer communities seriously. Unless we acknowledge and explicitly refer to the intimacy of life in hunter-gatherer settings with their multirelational and pluripresent dynamics we will neglect highly significant aspects of their being, she argues. This perspective also allows us to perceive the inhabitants as more than faceless human "stick figures", but instead as individuals with a diversity of roles and experiences. Seen from archaeology, where Big Data analyses and cross-regional approaches are prominent, it is refreshing to be reminded of the individual beings at the core of hunter-fisher-gatherer communities.

In response to the debate on scale amongst hunter-gatherers, Bird-David suggests the apparent conflict between arguments for small-scale and large-scale social worlds is linked to the different readerships addressed and the underlying research questions posed. The impact of dissimilar perspectives should not be underestimated. However, vague terminology adds to the confusion. What is implied by small-scale? How do we define and use the term? As Bird-David notes, the term has perhaps been used too readily in introductions to hunter-gatherers, with the risk of becoming descriptive rather than analytical. The term is clearly relative and could refer to quantitatively very different group sizes in different analytical contexts. Hence it should be explicated for all case studies, rather than employed as a self-explanatory concept. Bird-David does so predominantly through providing individual examples of relationships and interaction, focussing on the qualitative aspects of interaction rather than quantitative numbers, partly because she problematizes how we count members of a local group. Nevertheless, she accepts that providing demographic figures may be required for any wider comparison. This lack of agreement as to what constitutes a small-scale society is at the core of the present chapter, where Bird-David uses figures from Bird et al. (2019) to argue that despite numbers of "upwards of hundreds," the Martu community is still small-scale. If putting numbers on the table does not solve the issue, then we must perhaps return to the research questions: are we interested in quantifying the number of co-residents of an individual during their lifetime and the number of their personal contacts, or do we wish to explore qualitative aspects of interaction and their impact? The advantage of the first approach is of course that it will allow us to compare communities globally, while the latter may point to social behaviours and perceptions indicative of the scale of interaction as perceived by the community members themselves.

The existence of significant demographic and social diversity within extant and past hunter-gatherer communities is fully agreed upon in both archaeology and socio-cultural anthropology but may be under-communicated when seeking to describe similarities across the many different groups. The Nayaka and the Martu have very different historical trajectories and inhabit very different environments. While the Nayaka live in a region where agriculture and market-based economies have millennia-long histories, Western Australia was one of the last regions in the world to be impacted by colonialism. Similarly, the dense forest surrounding the Nayaka stands in great contrast to the landscape where the Martu live. While an argument can be made for both being involved in small-scale social networks, a quest for a mutually agreed term may distract us from the fact that the historical and environmental settings of two communities may in fact have resulted in quite distinct socio-cultural scales. The workshop challenged us to consider how they scale, how we scale and how scale matters. In a debate concerning scales within hunter-gatherer communities themselves and in academic analyses, the possibility of different perceptions of scale among hunter-gatherers such as the Nayaka and the Martu should not be forgotten.

Comment by Bram Tucker

The *Scale Matters* workshop was partially inspired by the apparently discordant claims, published within a few years of each other, that hunter-gatherers live in "nano-scale" societies (Bird-David 2017), or have large social networks (Bird et al. 2019). Within the first hour of our workshop, most of us became convinced that the two claims were largely in agreement. The apparent discord stemmed from different sub-disciplinary traditions, terminologies, and audiences.

As someone engaged with both the sociocultural and ecological-evolutionary approaches to hunter-gatherer studies, I have found the division between these approaches to be a constant source of frustration. The "sides" do not seem to read each other's work in sufficient detail to see the parallels and contradictions. Elsewhere (Tucker 2014) I have speculated about the origins of this division. Social and cultural anthropologists assume *à priori* that humans are social creatures who collectively imagine into existence diverse cultural worlds. Neodarwinian behavioral theory co-evolved with neoclassical economic theories of rational individualism. As a result, the two approaches find themselves on opposite sides of significant theoretical clefts: structure versus agency, and cultural relativism versus psychic unity.

Over the past two decades, theoretical and methodological advances have pushed the ecological-evolutionary approach closer to the sociocultural tradition (Fuentes 2004; 2016). Whereas twentieth century evolutionary theory emphasized inter-individual competition and explained away apparent altruism as self-interest-in-disguise, a growing number of twenty-first century scholars accept theories of cultural group selection by which one's group affiliations have an equal or greater influence one's fitness than individual traits and choices, so that people follow shared coordinative and cooperative norms even in the absence of individual advantage (Richerson and Boyd 2005). Thus, evolutionary anthropologists have arrived at the point where sociocultural anthropologists started, at the understanding that humans are social creatures in cultural worlds. Ethnographers with long-term fieldwork commitments such as Doug Bird, Rebecca Bliege Bird, and colleagues have learned that cosmological concepts like the Australian Dreamtime are inextricable from people's foraging behaviors and uses of resources.

That Bird-David, in her chapter in this volume, sees Bird et al.'s arguments about flexible group size and composition as a "strawman," illustrates just how far apart the sociocultural and ecological-evolutionary approaches to hunter-gatherer studies remain. Bird-David is correct, of course, that flexible group size and composition and the creation of kinship among non-biological relatives have been significant themes in social and cultural studies of foragers ever since the Man the Hunter Conference in 1968. But Bird et al. (2019) are correct that many paleoanthropologists and cognitive psychologists, particularly those working from non-human primate analogs and mathematical models, continue to assume that foragers, and humans generally, naturally assort by genetic kinship in hierarchically organized clusters. Bird et al.'s arguments might have been strawmen had they been published in *American Ethnologist*, but these arguments are not strawmen for the readers of the *Journal of Human Evolution*. Bird et al.'s article is a significant step toward closing the gap between approaches.

Bird-David argues in this volume that counting people is useful, but that it poses practical problems of who to count, and theoretical problems of whether a counted "group" compose a "society." Bird et al. agree. They use the concept of social networks to show that social structure transcends the small-scale of who one is spending time with at given moments in the day. Bird-David argues that "not-small" Martu social networks numbering "several hundred" are still "nano-scale" compared to nations. Ultimately, whether we call such grouping nano-, small-, large-, etc., depends on the comparisons we are interested in. Bird-David is defining hunter-gatherer scale in contrast to nations, whereas Bird et al. are defining scale among real-life hunter-gatherers in contrast to hypothetical "small-scale societies."

There remains a significant point of disagreement between Bird-David and Bird et al., and the fact that this point is not immediately obvious demonstrates how far we still have to go to bridge sub-disciplinary divides. It is nonsensical to ask whether "hunter-gatherers" live in "nano-" or "large-" scale societies, because hunter-gatherer is a scholar's category and not an objectively real thing. Group size probably predicts who we consider proper members of the hunter-gatherer category rather than the other way around. Notice that neither set of authors make comparisons to the 16th century Calusa of Florida, an example of a hunter-gatherer-fisher urbanized marine state (Thompson et al. 2018).

Indeed, neither Bird-David nor Bird et al. are actually arguing that being "hunter-gatherers" is the cause of social scale. Bird-David's discussion of scale among Nayaka is couched within a broader discussion of political encapsulation. Nayaka scale is small for social, historical, and political reasons. Bird et al.'s analysis of scale among Martu is framed around the significance of relational capital among semi-mobile people reliant on natural resources. Perhaps they are not talking about the same thing at all. Bird-David's arguments should be equally applicable to other minority indigenous communities regardless of their economic model, and Bird et al.'s arguments should apply to other mobile people in low-population density settings, including some farmers and herders.

References

Bird, Douglas W., Rebecca Bliege Bird, Brian F. Codding, and David W. Zeanah. 2019. Variability in the organization and size of hunter-gatherer groups: Foragers do not live in small-scale societies. *Journal of Human Evolution* 131: 96-108.

Bird-David, Nurit. 2017. Before Nation: Scale-Blind Anthropology and Foragers' Worlds of Relatives. *Current Anthropology* 58(2): 209-226.

Fuentes, Augustin. 2004. It's not all sex and violence: Integrated anthropology and the role of cooperation and social complexity in human evolution. *American Anthropologist* 106(4): 710-718.

— 2016. The extended evolutionary synthesis, ethnography, and the human niche: Toward an integrated anthropology. *Current Anthropology* 57:S13-S26.

Richerson, Peter J., and Robert Boyd. 2005. *Not by genes alone: How culture transformed human evolution.* Chicago: University of Chicago Press.

Thompson, Victor D., William H. Marquardy, Karen J. Walker, Amanda D. Roberts Thompson, and Lee A. Newsom. 2018. Collective action, state building, and the rise of the Calusa, southwest Florida, USA. *Journal of Anthropological Archaeology* 51: 28-44.

Tucker, Bram. 2014. Rationality and the green revolution. In *Applied Evolutionary Anthropology: Darwinian Approaches to Contemporary World Issues*. Mhairi Gibson and David Lawson (eds.). Pp. 15-38. New York: Springer.

What good is archaeology?
Archaeological and ethnographic scales

Robert L. Kelly

The Collaborative Research Center that facilitated the meeting on which this volume is based was titled "Culture-Environment Interaction and Human Mobility in the Late Quaternary." One project goal was to use ethnographic and ethnological data, as well as agent-based modelling to devise a model, a First African Frontier model, that accounts for how modern humans, some 50-75,000 years ago (or thereabouts), migrated out of Africa into Europe and, in fact, to the rest of the world. The original idea for this model was not archaeological but ultimately it must be tested against archaeological data.

This matters because the period in question, the late Pleistocene, during which modern humans expanded out of Africa, was a unique time in world history. The hunter-gatherers that we know today or from the recent past are firmly embedded in the landscape. They know their territories in minute detail. They have strong emotional ties to their lands places of stories, where the lives of ancestors are written into the landscape. Modern hunter-gatherers are also people who cannot move freely into new territories – because they are hemmed in by other groups, some hunter-gatherers, but most not. In contrast, the Pleistocene migration out of Africa entailed moving into some land occupied by other humans, notably Neanderthals and Denisovans, who were most likely living at very low population densities. And they also moved into land *not* already occupied by our genus: far eastern Russia, Australia, and, of primary concern here, the entire western hemisphere.

The movement from Africa into Europe, across Asia, and into the western hemisphere was, in geologic time, very fast, and entailed a level of migration, of territorial shift, quite unlike anything known among ethnographically known hunter-gatherers. In what ways do we expect these ancient hunter-gatherers to behave like those we know from ethnographic accounts and in what ways might we expect them to be different?

The problem is difficult, because archaeological and ethnographic data sources are not the same, and so analogies from ethnography are not easily carried over to the study of prehistory. I am speaking, of course, of the obvious fact that archaeologists cannot talk with the dead and cannot directly observe their practices and so we must test ideas with analyses of material culture, and yet much of ancient material culture has been lost to decay. Although our methods improve every year, today we cannot in most cases know with certainty the language people spoke, the particulars of their religion and cosmological beliefs, details of their kinship and social organization (although strontium analysis has allowed us to infer post-marital residence in cases), whether cross-cousins or parallel-cousins (or someone else) were preferred marriage partners, whether people thought of trees and stones and animals as "like persons," or all the other elements of human culture that helped structure what people did.

But I am also speaking of a great difference in scale that is the focal point of this volume. A long-term ethnographic study might be 50 years, and it might cover a country. But some archaeologists study human societies over enormous spans of time and over enormous spans of geography (Kelly 2016). Archaeology is good at seeking, analysing, and interpreting patterns in material remains over long spans of time and wide expanses of space. It is less good at consistently and systematically obtaining the minute detail that makes humanity interesting. Any effort to bring ethnological data and the enterprise of archaeology together must bear these two facts in mind and focus on archaeology's strength.

So, what do we do with the fact that archaeology cannot infer many of the elements of past human cultures that ethnographic research shows us matter, and that it operates with a different temporal scale? Answering this question requires us to think about two things: the scale of archaeological data and what investigative strategy best suits that scale.

The scale of archaeological data

Archaeological data have two essential elements: age and location. Archaeologists are compulsive about location; we try to record an object's provenience to the most precise level possible, using instruments such as an EDM (electronic distance measure) to record an artifact's location to ± 3 mm relative to a 3-dimensional grid system. But the artifacts in many sites have been moved, vertically and horizontally, through many processes making it difficult to associate even carefully plotted individual items with one another.

Age is also problematic. AMS radiocarbon dates come with standard errors in the 15 to 30-year range. That's excellent, but it means that the age's 95% confidence interval is 60-120 years – compare that to the standard length of a long-term ethnographic study. Worse, much of the time period of interest here in the Old World lies beyond the range of radiocarbon dating (~50,000 years). Sites more than 50,000 years old are dated by other, less precise means (e.g., optically-stimulated luminescence) that might provide a confidence interval of hundreds if not thousands of years. And this means that what archaeologists might consider a tightly-dated archaeological assemblage is a potential aggregate of artifacts left behind by many individuals – men, women, children, the elderly, etc. – during possibly many different uses of a location.

The temporal scale of archaeological data, even under the best of circumstances, is obviously quite different from that of ethnographic data. We must consider this when asking, what can archaeology tell us? What about human society and culture can we infer from those artifacts that survived what Francis Bacon called "the shipwreck of time" and that come from a record that is a palimpsest of the evidence of many activities?

We can draw an analogy between archaeological data and a radio. Sometimes the radio signal comes in clear, but sometimes it is poor, and full of static. At those times, one might be able to discern that the voice is male, and speaking English, but the precise words are impossible to hear. This does not mean the words are unimportant, only that we cannot hear them. If we cannot hear the "words" of prehistory, then we can either abandon archaeology or decide to use what it can consistently provide, its "strong signal".

A research strategy

Let me be blunt: the temporal and spatial patterns uncovered by archaeology, especially of the time period of concern here, most likely reflect, at archaeology's temporal scale, the broad ecological, subsistence, and demographic conditions of life. It is these factors that provide the "strong signal" of archaeology and thus its first-order interpretations. And let me be clear: Archaeology does sometimes allow glimpses at finer scales (as with Ötzi, the Neolithic man found frozen in the Italian Alps), but the most assured things we can *systematically* infer, and that provide us with an important comparative base, are the ones that reflect the elements of life that deal with food, security, and reproduction.

I say this because numerous cross-cultural ethnological studies show that an environment's ecology exerts a strong influence on hunter-gatherers (Kelly 2013; Binford 2001). And anyone who accepts global warming as a reality, and something we must adapt to, must also accept that people before us have had to contend with climate change.

Humans also have daily caloric needs – varying with age, gender, size, and workload – and if people cannot satisfy those needs then little else can follow because those people will be dead. Finding food is basic. (We tend to forget this since we live in a world where the fortunate among us do not have to worry where their next meal is coming from.) The environment sets potentials and limits to ways of satisfying that daily need. We could begin with the banal fact that foragers will not eat much plant food in the arctic and then move to the less banal fact that the abundance and distribution of game and plants, combined with their costs of acquisition and caloric value, will condition which foods are used (claims that can be verified or not using the plant and animal remains recovered in sites). Likewise, the abundance and distribution of sites of a given time period are first and foremost telling us something about the abundance and distribution of people.

Given where I think the first-order interpretations of archaeological data lie, I suggest that the often-disparaged optimal foraging models offer a useful research strategy to approach interpreting the archaeological record. Optimal foraging models were brought into anthropology from the field of evolutionary ecology, where they were intended to unify ecological approaches with an evolutionary perspective. These models were brought to anthropology by Bruce Winterhalder and Eric Smith, both of whom studied hunting and gathering cultures. It is probably this historical accident, rather than the reprehensible assumption that hunter-gatherers are "closer to nature", that is responsible for their common use in the study of hunter-gatherers. All humans are equally "close to nature" and equally not.

In anthropology, the approach of evolutionary ecology takes the name *human behavioral ecology* (HBE) and modifies the models to account for the particulars of humans (e.g., division of labour, central place foraging, environmental knowledge, symbolic labelling of food and activities). These models privilege material conditions, especially food and reproduction, and focus on "maximizing" behaviours (e.g., how does an organism maximize reproductive advantage under such-and-such conditions?). Many anthropologists reject these models, claiming they are nothing more than sociobiology (which they link to racist views), or capitalism written into the natural world. And yet, the models have been tested against ethnographic case studies and proven to be useful in predicting human behavior (Kelly 2013).

However, there is a scale problem here as well. Foraging models were developed to model individual decision-making, moment-by-moment. Given conditions A, B, C ... what food resources might we expect the individual forager on a daily foraging trip to collect from an environment? (Or it could be what resources do we expect to be shared, or what size group do we expect an individual to opt to live in, and so on.) But archaeology, as I pointed out, deals with palimpsests that include the material evidence of human behaviour over a long time span but almost never that of a single individual's daily decisions. If a diet is broadening and contracting over some interval of time, we will not see that in an aggregated dataset – all the animal and plant remains that provide evidence of diet might be combined into an assemblage that cannot be disaggregated. This only means, however, that we must evaluate the data with the recognition in mind that, in this example, we are looking at the maximal diet breadth. And it means that a significant change in diet breadth between, say, two time periods indeed reflects a significant change in human behaviour. And the longer the time period entailed in formation of archaeological assemblages, the greater the likelihood that the "strong signal" in those data will reflect inescapable realities of foraging lifeways, and the lower the likelihood that other cultural variables produce significant patterning in the data *over the reaches of time that archaeologists normally must confront*. Again, I do not mean that cultural ideas have no effect. But cultural knowledge can change rapidly, and probably did change during any archaeologically-defined Palaeolithic period (e.g., the Aurignacian). And this suggests that not all changes in, say, a people's definition of relatives or cosmological beliefs, will lead to a large-scale change in, e.g., subsistence, especially if that change correlates with something that could affect subsistence choice, such as a change in climate or population density.

I admit this approach could lead us astray. But the utility of HBE foraging models is that they provide a way to know when an idea is wrong. Take a simple example: what if some foods are tabooed, or some taken for non-food reasons (e.g., certain plants for medicinal needs)? An optimal foraging model might simply say: if foragers are behaving according to a certain set of principles grounded in ecological and evolutionary theory, then in a particular environment we expect them to take, as food, resources A through G. If the archaeological record shows something different (resource E is not taken, and instead resource H shows up), then we have good grounds on which to argue that some factor is at work other than those incorporated into the model. HBE's foraging models provide a useful strategy because they can recognize information that qualifies or even negates the original assumption.

A useful strategy, one that can help tell us when we are wrong, is important because it is, of course, absolutely true that humans live in a culturally constructed world. We deem some foods to be edible and others to be inedible; some people are proscribed as mates, and others are prescribed. We treat the environment one way if we think that trees, animals, rivers, and rocks are ancestral spirits and another way if we think God has given us a mandate to dominate the earth. HBE's models are not ready-made answers, but they provide a research strategy that is suited to the large temporal and geographical scales of palaeolithic archaeology.

Now let me turn to the last part of the first African Frontier, the colonization of the western hemisphere, to hypothesize what the nature of that frontier might have been like.

Colonization of the New World

The western hemisphere and Australia, as well as portions of far northern Asia, were lands first occupied by modern humans. They are particularly interesting cases since they were, as far as humans were concerned, *terra incognita* in the late Pleistocene.

I will focus on the western hemisphere as that is the case I know best. The timing, route, and adaptation of this region's first inhabitants are highly contentious topics. I can only give a quick summary here since my point is to discuss scale issues as they relate to hunter-gatherer migration. I currently think the best evidence points to an entry between 14,500 and 16,000 years ago to the continental US. There are a few very early (>20,000 cal BP) sites (e.g., the Cerutti Mastodon site in California, Chiquihuite and Coxcatlan Caves in central Mexico, and the White Sands footprints in New Mexico (Ardelean et al. 2020; Bennett et al. 2021; Holen et al. 2017; Somerville et al. 2021), but these have convinced few archaeologists (Braje et al. 2017; Chatters et al. 2021; Potter et al. 2018; 2021). Better evidence comes from the ~16,000-year-old Gault/Friedkin (Texas) and Coopers Ferry (Idaho) sites (Davis et al. 2019; Waters et al. 2011; Williams et al. 2020), and the 14,500-year-old sites of Page-Ladson (Florida) and Paisley Cave (Oregon; Halligan et al. 2016; Shillito et al. 2020).

The route from Asia could have been along the western, largely ice-bound coast or through the ice-free corridor (Potter et al. 2018); I lean toward the coastal route, where the earliest sites are located (McLaren et al. 2018). Boats might have been employed, though they were likely modest forms, and not ocean-traversing vessels. Slightly later populations might have come through the ice-free corridor once plant and, especially, animal food became available there.

But let me turn to the nature of adaptation at the time of colonization. The earliest culture that we know of in North America is the *Clovis* complex. Its primary material hallmark is a large, lanceolate projectile point with basal "flutes" created by one or more flakes removed from the base on each side, accompanied by grinding of the base's edges. These appear in all 48 states of the continental US, and a few occur in Canada, Alaska, and Mexico. The tradition might continue into South America in the form of (sometimes) fluted "fish-tail" projectile points.

Clovis currently dates to 13,050 to 12,750 cal BP (Waters, Stafford, and Carlson 2020), but this range is based on fewer than a dozen dated sites, which are clustered in the Plains and the northeast. Statistical studies show that the first appearance of Clovis is earlier than it appears, perhaps as early as 14,500 years ago (Prasciunas and Surovell 2014). It's likely that Clovis appeared first in the far west (assuming the coastal migration route is correct), where it has defied efforts to date it. Thus, our dated sample is both small and geographically biased. Regardless of whether someone first set foot on the continent south of the ice sheets 14, 15, or 16,000 years ago, it appears that virtually all of the western hemisphere was occupied in a short period of time. Why?

In 1988, Lawrence Todd and I proposed one model (Kelly and Todd 1988). We pointed out that the first entrants to the New World would have been hunters, since they were coming from the arctic (I think this would have been true even if they used a coastal adaptation since arctic coastal peoples are also terrestrial mammal hunters). This means they were comfortable with game, but perhaps less so with plants. Since mammal anatomy is basically the same – mammoths are just scaled-up rabbits – the knowledge of preparing meat can be transferred across environments; this is less true of plants. While hunting benefits from local knowledge of animal behaviour and terrain, it is possible to survive by hunting in unfamiliar land: 17th through 19th century European fur trappers lived primarily off game as they moved across the North

American continent (Hudson's Bay Company policy required trappers to live off the land or eat whatever they could get in trade with Indigenous peoples).

Animals are available year-round (although they are in better condition at some times than at others), while plants are not. Plants, in addition, can have more time-consuming processing needs and some, such as acorns, are quite labour-intensive and require figuring out how to use them (e.g., acorns are full of tannic acid, and require pounding and leaching to remove it). In fact, plants were probably not an important part of diet as the tools for their processing, such as grinding stones, show up a few thousand years after Clovis. The few traces of plant foods recovered from Clovis-age sites are mostly snack foods, such as berries (this is likely not a function of preservation; Kitchel and Mackie 2022).

So we proposed that arctic hunters, after entering North America south of the ice sheets would have continued their arctic adaptation and focused on hunting. This would have been a viable adaptation in the terminal Pleistocene when North America contained a variety of large game (which soon became extinct, possibly due to human predation but we will leave aside that contentious issue). We have solid evidence that Clovis hunters took mammoths, mastodons, and bison; and indirect evidence they took horses, camels, and sloths.

Clovis hunters would have found themselves in a world of naïve game, animals who had never experienced human hunters before. There are not many parallel cases, but there are some in which human hunters (or wolves reintroduced to Yellowstone National Park) experienced naïve game (Kelly and Prasciunas 2007). In these cases, the kill rate is very high. But the animals respond and within a few years learn to avoid the new threat. In a world where there are no other humans beyond the colonization front, hunters would know that they could do better, i.e., achieve a higher return rate, if they simply moved to new territory.

The catch-22 is that those hunters did not learn about their current environment's unique properties, including the location, seasonal timing, and processing needs of plant foods. Hunting allowed them to move into new environments, but the high return rates of naïve, non-depleted herd animals coupled with the availability of land devoid of humans and populated by naïve game also encouraged Clovis people to keep moving into new environments across North America. The implication, of course, is that the same cultural group carried Clovis technology across the continent.

Foraging models support this reconstruction. Without going into detail (see Kelly 2013), a base model predicts that hunter-gatherers will move before exploiting all foods that are economically obtainable from a settlement. For example, if one could forage at an economic gain up to 6 km, then the camp might move after using food within 3 km of camp (assuming Binford's [1980] "half-radius" foraging area; this follows what is known as the "marginal value theorem", which has been ethnographically demonstrated [Venkataraman et al. 2017]). However, a model in which the return rate declined as a function of occupation of a camp, e.g., under conditions where animals respond to hunters and make themselves more time-consuming to hunt, encouraged movement at an even shorter distance. In other words, we expect a colonizing population dependent on hunting to move quickly across a landscape populated by naïve prey.

The theoretical model focused on daily behavior. We "upscaled" the model to inform us about movements of territory, something that happened perhaps every few years; that is, we treated a territory as if it were a camp. Is this a proper thing to do? That's where archaeology provides a test. And unfortunately, no one has yet made that test as it requires copious dates on sites across the continent. However, an alternative is that migration was driven solely by population growth, with daughter groups moving away when local carrying capacity was reached. Modelling suggests this would require a population growth rate far above that recorded for prehistoric hunter-gatherers, which hover around 0.04%, not the higher rates observed among ethnographically-known hunter-gatherers (Zahid et al. 2016; Prasciunas and Surovell 2014). Consequently, the movement was probably not driven by population growth alone.

The geographic scale of social relations

Let me return to another question of concern, namely, the scale of social relations because our reconstruction implies widely scattered Clovis residential groups. Ethnographic data show that nomadic hunter-gatherers live most of the year in groups of about 25 people, with short-term seasonal gatherings of larger groups. The figure, 25, seems to be true almost regardless of the environment (Hamilton et al. 2007). Bruce Winterhalder (1986) showed that it appears to balance depletion of the local foraging area with the need to have enough hunters in a group so that someone is successful (at large game hunting) at a rate that will feed the group (see Kelly 2013). But 25 people is too small to maintain reproductive viability. Under conditions of extremely low population density, which was the case for the population that colonized the New World, a residential group had to stay in touch with enough other widely spaced residential groups to avoid extinction. Is there any evidence of these larger groups during the colonization of the North America?

This is a tough question. For later time periods, archaeologists conduct social network analysis with ceramics as the basic data. For the Clovis time period, we mostly have stone tools; network analysis can be applied to the lithic raw materials from which those tools were fashioned. Doing so with Clovis assemblages, Buchanan et al. (2015) found three major networks: one in the northeast that stretches from west of the Great Lakes, south to Missouri, over to South Carolina and north to Maine. A second covers Texas and the southern and central Plains, and a third covers the northern Rocky Mountains out to the western north plains. These regions receive some support by concomitant regional differences in Clovis projectile point form as well (Buchanan et al. 2014).

I do not mean to imply that these (large) regions were ones in which every individual interacted with every other person. They might, however, be regions where individuals were more socially connected, in a spider-web way, than they were between regions; thus, they tell us something about the geographic scale of social connections during colonization. And the message here is that these groups of 25 or so foragers were each embedded in a geographically large social network, as Bird et al. (2019) point out for modern Aboriginal Australians. But let's not trade one stereotype for another: it is clear that as population grew in North America, and economic organization shifted, in many places from hunting and gathering to agriculture, that the geographic extent of groups – in other words, their scale – also changed. The geographic scale and social content of groups are part of the adaptive process. We should not expect them to be always the same everywhere.

Men and women

Even with a gross temporal and spatial scale, can archaeology inform us about such things as the division of labor? Ethnographic data support a division of labour, regardless of environment, in which men hunt and women gather. This is most likely due to the need to breastfeed children – sometimes for several years – and consequently for mothers to keep children with them.

Since childcare is more compatible with plant gathering than hunting, women gather (see discussion in Kelly 2013).

However, it's possible that a hunting-focused adaptation south of the ice sheets produced different relations between men and women. Haas et al. (2020) argue that the number of hunters in early western hemisphere foraging groups might have consisted of nearly equal numbers of men and women. The analysis is difficult as it relies on a very small sample of human burials, as well as on the interpretation of grave goods (which is a fraught exercise, given the highly symbolic nature of burial ritual). Nonetheless, using the presence of hunting equipment as a guide to what people did in life, Haas suggested that 30-50% of women were likely to have been hunters.

Archaeologists have long denigrated a reconstruction of Clovis life that portrayed men out hunting mammoths while women sat at camp, breastfeeding children. Women were no doubt doing many other things. In fact, using ethnographic data, Waguespack (2005) found that women in foraging societies do an increasingly larger percentage of tasks other than direct food acquisition as the percentage of meat in the diet increases. These tasks include childcare, but also firewood collection, and especially clothing manufacture. But the ethnographic dataset of meat-dependent groups is biased toward arctic peoples, i.e., groups that have significant clothing requirements. What would women do in a hunting society of say, the southeast US, with much lower clothing requirements? One guess is that they could be incorporated into the hunt, and shift a group's tactics from individual stalking, or sit-and-wait hunting, to communal hunting, which might have increased the per capita return rate. In these cases, women (childless, past reproductive age, or able to leave an infant in camp in the care of another) could have been armed with atlatls and used to drive and even kill game. We do not know the answer yet. But I would not expect a hunting-dependent culture outside the arctic to look exactly like arctic cultures.

The First African Frontier

The first African frontier in and outside of Africa would have been similar and different to the North American case. Those intrepid hunter-gatherers who ventured out of Africa would not have been coming from the arctic, quite the opposite in fact, and so they may have been less focused on game. Those foragers were also probably not entering land with naïve fauna since Europe and Asia were already occupied by Neanderthals and Denisovans. There is also no grand barrier to movement like the ice sheets in North America that might have made "rearward" movements more difficult.

On the other hand, evidence for intensive plant food processing also appears late in the Old World, and all optimal foraging models place game animals as the top-ranked foods (balancing search and harvest/processing costs with the calories acquired). Those moving out of Africa were probably hunting-focused, though maybe not to the extent of Clovis hunters. This suggests that the migratory wave out of Africa was driven more by population growth and subsequent (slow) territorial depletion, than by the attraction of higher hunting return rates in an unpopulated region with naïve fauna. This is a process that could be modelled using, e.g., the approach recently employed by Klein et al. (2021). Combined with an environmentally informed "ideal-free distribution" foraging model, we could also predict which regions would be occupied first, second, and third. This does not mean the model is right, but, tested against archaeological data, it provides us with knowledge of when we are wrong, when factors other than the simple ones entailed in foraging models are not driving the pattern as revealed by archaeology's "strong signal".

Conclusion

Scale is a fact of social life; it is also a fact of research, driven largely by the nature of our data. The large-scale patterns revealed by archaeology, its "strong signal", are most likely revealing issues of ecology, human subsistence, and reproduction. The patterns we observe in the archaeology associated with the movement of palaeolithic hunter-gatherers across a continent is most likely to be explained by those factors communicated in archaeology's strong signal. This does not mean that other factors were not at work, only that they are difficult to discern at the scale palaeolithic archaeology can record information. Nonetheless, the interpretation of large-scale patterns uncovered by archaeology are complemented and potentially tested by small-scale studies at those archaeological localities amenable to research that retrieves fine-grained information, something more than just the "strong signal". They can also be hypothesized through agent-based models incorporating social variables (see Widlok and Henn, this volume). I think HBE is a useful research strategy because it is suited to large-scale archaeological data and, by making archaeologically testable propositions, provides a way to know when it is leading us astray. There is not much more we could ask for in a research strategy.

References

Ardelean, Ciprian F., et al. 2020. Evidence of human occupation in Mexico around the Last Glacial Maximum. *Nature* (7/22/2020): DOI: https://www.nature.com/articles/s41586-020-2509-0.

Bennett, Matthew R., et al. 2021. Evidence of humans in North America during the last glacial maximum. *Science* 373: 1528-1531.

Binford, Lewis R. 1980. Willow smoke and dogs' tails: hunter-gatherer settlement systems and archaeological site formation. *American Antiquity* 45: 4-20.

Binford, Lewis R. 2001. *Constructing Frames of Reference*. Berkeley: University of California Press.

Bird, Douglas, et al. 2019. Variability in the organization and size of hunter-gatherer groups: foragers do not live in small-scale societies. *Journal of Human Evolution* 131: 96-108.

Braje, Todd J., et al. 2017. Were Hominins in California ~130,000 Years Ago? *Paleoamerica* 3: 1-3.

Buchanan, Briggs, et al. 2015. Lithic networks reveal early regionalization in late Pleistocene North America. *Journal of Archaeological Science* 65: 114-121.

Buchanan, Briggs, Michael J. O'Brien, and Michael Collard. 2014. Continent-wide or region-specific? A geometric morphometrics-based assessment of variation in Clovis point shape. *Archaeological and Anthropological Sciences* 6: 145-162.

Chatters, James C., et al. 2021. Evaluating claims of early human occupation at Chiquihuite Cave, Mexico. *PaleoAmerica* DOI: 10.1080/20555563.2021.1940441.

Davis, Loren G., et al. 2019. Late Upper Paleolithic occupation at Cooper's Ferry, Idaho, USA, ~16,000 years ago. *Science* 365: 891-897.

Haas, Randall, et al. 2020. Female hunters of the early Americas. *Science Advances* 2020; 6:eabd0310.

Halligan, Jessi J., et al. 2016. Pre-Clovis occupation 14,550 years ago at the Page-Ladson site, Florida, and the peopling of the Americas. *Science Advances* 2(5), p.e1600375.

Hamilton, Marcus J., et al. 2007. Nonlinear scaling of space use in human hunter–gatherers. *Proceedings of the National Academy of Sciences* 104: 4765-4769.

Holen, Steven R., et al. 2017. A 130,000-year-old archaeological site in southern California, USA. *Nature* 544: 479-483.

Kelly, Robert L. 2013. *The Lifeways of Hunter-Gatherers*. Cambridge: Cambridge University Press.

Kelly, Robert L. 2016. *The Fifth Beginning: What Six Million Years of Human History Can Tell Us about Our Future*. Berkeley: University of California Press.

Kelly, Robert L., and Lawrence C. Todd. 1988. Coming into the country: early paleoindian hunting and mobility. *American Antiquity* 53: 231-244.

Kelly, Robert L., and Mary M. Prasciunas. 2007. Did the ancestors of Native Americans cause animal extinctions in Late Pleistocene North America? In *Native Americans and the Environment: Perspectives on the Ecological Indian*. M.E. Harkin and D.R. Lewis (eds.). Pp. 95-122. Lincoln, University of Nebraska Press.

Kitchel, Nathaniel R., and Madeline E. Mackie. 2022. Plants and subsistence during the fluted-point period of the northeast. *American Antiquity* 87, in press.

Klein, Konstantin, et al. 2021. Human existence potential in Europe during the Last Glacial Maximum. *Quaternary International* 581-582: 7-27.

McLaren, Duncan, et al. 2018. Terminal Plestocene epoch human footprints from the Pacifc coast of Canada. *PLoS ONE* 13(3): e0193522

Potter, Ben A. et al. 2018. Arrival Routes of First Americans Uncertain. *Science* 359: 1224.

Potter, Ben A. et al. 2021. Current Understanding of the Earliest Human Occupations in the Americas: Evaluation of Becerra-Valdivia and Higham. 2020. *PaleoAmerica*. DOI: 10.1080/20555563.2021.1978721.

Prasciunas, Mary M., and Todd Surovell. 2014. Reevaluating the duration of Clovis: the problem of non-representative radiocarbon. In *On the Edge of a New Understanding*. Ashley M. Smallwood and Thomas A. Jennings (eds.). Pp. 21-35. College Station: Texas A&M Press.

Shillito, Lisa-Marie, et al. 2020. Pre-Clovis occupation of the Americas identified by human fecal biomarkers in coprolites from Paisley Caves, Oregon. *Science Advances* 6: eaba6404.

Somerville, Andrew D., Isabel Casar, and Joaquin Arroyo-Cabrales. 2021. New AMS radiocarbon ages from the preceramic levels of Coxcatlan Cave, Puebla, Mexico: A Pleistocene occupation of the Tehuacan Valley? *Latin American Antiquity* 32: 612-626.

Venkataraman, Vivek V., et al. 2017. Hunter-gatherer residential mobility and the marginal value of rainforest patches. *Proceedings of the National Academy of Sciences* 114: 3097-3102. https://www.pnas.org/cgi/doi/10.1073/pnas.1617542114.

Waguespack, Nicole. 2005. The organization of male and female labor in foraging societies: implications for early paleoindian archaeology. *American Anthropologist* 107: 666-676.

Waters, Michael R., et al. 2011. The Buttermilk Creek complex and the origins of Clovis at the Debra L. Friedkin site, Texas. *Science* 331, 1599e1603.

Waters, Michael R., Thomas Stafford, Jr., and David L. Carlson. 2020. The age of Clovis – 13,050 to 12,750 cal yr B.P. *Science Advances* 6, DOI: 10.1126/aciadv.aaz0455.

Williams, Thomas J., et al. 2018. Evidence of an early projectile point technology in North America at the Gault Site, Texas, USA. *Science Advances* 2018; 4: eaar5954.

Winterhalder, Bruce. 1986. Diet choice, risk, and food sharing in a stochastic environment. *Journal of Anthropological Archaeology* 5: 369-92.

Zahid, H. Jabran, Erick Robinson, and Robert L. Kelly. 2016. Agriculture, population growth and statistical analysis of the radiocarbon record. *Proceedings of the National Academy of Sciences* 113: 931-935.

Comment by Graeme Warren

Small town England in the early 1980s had an overriding aroma of cheese and onion crisps, damp concrete, and failure. For the young teenage me, the bold, brash American Football that was being shown on a new once-weekly TV show on Channel Four was a bright and shiny escape from this grey world. I became mildly obsessed. Aside from Channel Four, American Football was barely covered in the UK media, but, I had one other form of access, an unexpected benefit of living near to the Cold War era American cruise-missile base at Greenham Common. On a Sunday evening I could, just about, pick up the radio broadcasts of the American Armed Forces Network from the base. Live commentary on games! The reception was very poor, fading in and out. The presenters used unfamiliar technical terminology, and many cultural references sailed over my head. But with persistence, some inference and lots of learning, I could listen to the radio broadcasts and keep track of the games. Making sense of these exotic, uneven and inconsistent broadcasts was even enjoyable.

I tell this story, of course, because of the analogy made to the radio in Robert Kelly's paper 'What Good is Archaeology?'. This is a stimulating short paper, and there is much that could be discussed about many aspects of it. But in keeping with the aims of this volume, I will focus my response on scale.

Kelly argues that whilst the radio signal of prehistory is sometimes clear, it is more frequently 'poor and full of static'. Because of this, we should develop strategies to use this static as our archaeological 'strong signal'. At times, Kelly argues that we might be able to discern a male, English speaking voice on the radio. But our focus should not be these moments because they don't offer any *systematic* data. Instead, we should work out how to engage with the static. What a curious way of listening to the radio!

As I argue in my paper in this volume, archaeology is characterised by mulit-scalar temporal data. Sometimes we have 'static' – to stick with the radio analogy. But sometimes we can make out the gender of the speaker, and the language they are communicating in. This is non-trivial. It matters if the speaker is communicating in English, French, Somalian, Mandarin or a Khoisan language. It matters if they are male, female or neither. Kelly and I are both male and we speak English. But if you listen to our voices you would also identify further differences between us. We should not dismiss these kind of insights.

Archaeological data also involves moments of sharper chronological resolution – moments where, perhaps, when we can hear the words of the broadcasts. In this regard, an array of new analytical techniques are enhancing our ability to provide details, and refining the resolution available to us. Bayesian modelling of radiocarbon dates, for example, means that rather than working on a 60-120 year resolution as claimed by Kelly, we can sometimes approximate to generational time frames. This puts us within a similar time scale to some long-term ethnographic studies.

Kelly suggests that archaeologists can't identify 'details of kinship' and preferred marriage partners. But this is not true. Kinship is increasingly something we do discuss. Recent genomic analysis of individuals buried in the Early Neoltihic Hazelton North Chambered tomb in southern England (Fowler et al. 2021), for example, shows that patrilineal descent appears to have been a key determinant of inclusion in the tomb, but location with the chambers was determined by female descent. Step-sons appear to have been incorporated into the lineage.

These finer grained aspects of our data matter. They matter because such details offer points of engagement with the humanity of the past. But they also matter because of how the archaeological record is formed. Numerous studies have demonstrated that hunter-gatherer groups create, use and deposit material culture according to specific ways of understanding the world: Peter Jordan, for example, has highlighted that the deposition of animal remains by the Khanty is spatially complex and bound up with negotiations between hunters and animal spirits (2003). We need detail to reconstruct these cultural practices and to understand how our sites are formed.

In this context, an archaeological retreat to the largest of scales and the claim that the abundance and distribution of sites (or animal or plant remains) is 'first and foremost telling us something about the abundance and distribution of people' is missing a key step. The abundance and distribtion of sites is, first and foremost, telling us something about the formation of the archaeological record. This arises from an interplay of activities in the past and in the present which create the material evidence we work with. Making sense of this record requires that we are sensitive to the multiple scales of our data: the smaller scale is our only hope of understanding how practices in the past influenced site formation, which is the basis of the evidence which we can use at larger scales.

The movement to the large scale in archaeology has been fed by many factors. In part, this is a reflection of the nature of (some of) our data, but the increasing availability of big data, the processing power to deal with it, and broader trends in academic publishing are also part of this trend. We should be careful what we lose in pursuing it (Cunningham & MacEacharan 2016). The large scale and the long term isn't the only thing that 'Archaeology is Good For'. Choosing to listen to the static and not the other parts of the archaeological 'radio signal' feels like a counsel of despair. Kelly argues that archaeology is "less good at consistently and systematically obtaining the minute detail that makes humanity interesting". That is not a description of archaeology that I recognise. It does not match my experience of working with students and volunteers and seeing their reaction to the recovery of a lithic or pot sherd. The excitement at a moment of connection with the past is a moment of attenuated interest in the character of a material, and the lives it was once bound with. Archaeology is much more than just static. And uneven quality radio signals can still be listened to. How else could I keep track of the scores?

References

Cunningham, Jerimy J., and Scott MacEachern. 2016. Ethnoarchaeology as slow science. *World Archaeology* 48 (5): 628-641.

Comment by Brian Codding

In this provocative essay, Kelly asks "what good is archaeology?" To be sure, as a collection of methods applied to the study human behavior, archaeology is limited. Specifically, Kelly notes that "archaeology cannot infer many of the abstract elements of past human cultures that ethnographic research shows us matter…" Given this, is it "possible to get anything of value from archaeology at all?"

Central here is a recognition of what we can, and cannot, "see" with archaeological data. Archaeology cannot "see" the scale of past social organization with the precision of an ethnographer, but it is equally sure that there are aspects of social organization in view. Acknowledging this limitation is a first step in identifying what good is archaeology.

This is in some ways reminiscent of Christopher Hawkes' 1954 essay on "Archaeological Theory and Method: Some Suggestions from the Old World" featured in American Anthropologist (Hawkes 1954), in which he outlines his (in)famous "ladder of inference". The rungs on the metaphorical ladder move from the observed unit of study, material remains, upward to those aspects of human society most removed from material objects, such as religious beliefs. With each rung, the archaeologist makes an inferential leap, creating less and less certainty about the claims being made. With this ladder in mind, we too can focus on what archaeology can do well, and what it should perhaps leave to the ethnographers.

However, moving beyond identifying limitations, Kelly highlights how we can bolster our inferences as we climb Hawkes' rungs by leveraging theory, specifically, theory from ecology. As Kelly notes, "ecology must exert a strong influence on hunter-gatherers" (and on post-industrial society as well, as is clear from global climate change, and the current pandemic). Using this fact and theory designed to amplify it, Kelly argues that we can help resolve some of the issues with archaeological data.

Leveraging these tools, Kelly examines how high we can reliably climb on Hawkes' ladder to understand social patterning among early colonizers of the Americas, even from course grained archaeological data. For examples, he reviews how the general approach has elucidated the geographical scale of past social interaction spheres, and the relative dietary contributions of women and men.

This reflexive turn rewards the reader with insights about how we can best examine the scales of hunter-gatherer social organization in the past, inferred from the material remains they left behind. In so doing, Kelly illustrates the good archaeology can provide.

References

Hawkes, Christopher. 1954. Archeological Theory and Method: Some suggestions from the Old World. *American Anthropologist*, 56(2), 155-168.

Upscaling forager mobility and broadening forager relations

Thomas Widlok and Stephan Henn

Introduction: The problem of upscaling local mobility

Ethnography has taught us to mistrust models of mobility that are purely resource-driven: We know that foragers often move for social reasons and rather anchor their moves in a region than to randomly move all over the place driven by ecological necessity. But the ethnographic models of mobility also have their limitations since they beg the question as to how things change in terms of long-term processes such as human movement out of Africa. Moreover, the ethnographic models tend to create a fundamental rift and bipolar opposition between small-scale foragers and larger populations with little indication as to how one could transform into the other.

In this chapter we report on some results from agent-based simulations that allow us to envisage a move from local to cross-regional mobility and a shift from small-scale mobility to larger-scale dispersal. We revisit the fission-fusion pattern and suggest how it can be reconceived so that it connects to out-of-region migration. We simulated how kinship rules influence population size and suggest how such simulations can help us explain moves from small to larger and from latent to actual wider networks. Finally, we discuss how hunter-gatherer ways of perceiving their environment and their social relationships can be reconciled with scenarios in which hunter-gatherers can upscale their mobility and broaden their social relationships without a categorical break with their modes of perception.

The collaborative research centre from which this contribution arises was entitled "Our way to Europe" and several case studies were conducted on present-day hunter-gatherers (in central African rainforest environment, in east African savanna environments and in southern African semi-desert environments[1]). These case studies strengthen the ethnographic record of human mobility which also serves as a baseline when modelling mobility of the past. However, to simply take cases of present-day African hunter-gatherers as analogues for the past has its limits since these groups are in fact people who did not move to Europe and have shown no inclination to do so. Their mobility is not one of inter-continental travel, it is not even one of far-distance and cross-regional migration. At the same time, when seeking to make the ethnography of contemporary people productive for questions of long-term dispersal, it seems likely that the evidence from today's African hunter-gatherers and their mobility patterns may be more relevant to processes in the distant past than the typical intercontinental migration of today which is conditioned by nation states and modern travel infrastructures. What would anthropological models of African foragers need to look like that try to explain how a large-scale intercontinental move emerged against the backdrop of the mobility patterns that we find wide-spread in small-scale societies that are observed today? How did a movement to Europe emerge from societies that were mobile but not migratory in the narrow recent sense?

The conventional models of fission-fusion dynamics

Why do we need "special" models to deal with human expansion out of Africa in the first place? After all, there are highly technical general models of mobility in existence (see Widlok 2016, 2017a). Most of these are rational choice models which claim wide applicability across time and space. For hunter-gatherer studies Optimal Foraging Theory (OFT) is a well-known example of these types of models (Winterhalder & Smith 2000). OFT has been used to model not only forager mobility across the world and back deep into time (see Kelly 1995), but they have also been employed to very different forms of mobility, for instance the movements of visitors who "appropriate" an exhibition space in a museum (Widlok and Eghbal-Azar 2012).

Most natural scientists implicitly or explicitly adhere to such rational choice models which they often consider to be "culture-free" as these models claim to tap into universal rational strategies such as "least effort for maximal returns". However, there are good reasons to question whether these rational choice models are indeed universal or culture-free. To begin with, these models have emerged in a particular cultural situation, enlightenment

1 See https://www.sfb806.uni-koeln.de/index.php/projects/s-supraregional-systems/e3

Europe, which was predicated economically and politically on an expansionist, imperial and growth-oriented culture. The "frontier" notion of humans constantly striving for better living space has its origin in the colonization of America, in particular the American West, which is a very specific cultural situation (see Brody 2001). In this case the people moving were the desperate poor, exported or fled from Europe, with no home to return to, with not much to lose, with no particular attachment to the new land they had come to live on, with contempt towards the indigenous people already living there and at least in part driven by an ideology of superiority, of "multiply and occupy" (see Turner 1893 for details on the American frontier). It is these very specific cultural conditions – repeated by settler communities in other places, notably Australia and southern Africa – that brought about the rationalist "optimal" foraging paradigm: The insecure existence of the frontier explorers depended on their being able to occupy "free" land wherever they found it. It also depended on their ruthlessness to exterminate or expunge any locals, and on their determination to move always forward and never backward. Not only was this group subscribing to an ideology that makes them search for "ever greener pastures" elsewhere, what Brody (2001) has identified as a general agriculturalist bias towards the world. Treating land and space as a resource has subsequently been amplified considerably by "the great transformation" that made land an item of markets and capital investments, disembedded from its previous social institutions and cultural meanings (see Polanyi 1944). The explorer-colonizer-capitalists readily exchange one place to stay for another one if the opportunity arises. This is still true for many descendants of European settlers in Africa and elsewhere up until today. Our research in Namibia shows that indigenous Africans often tend to cling to a particular piece of land although they find it difficult to make ends meet living off that land, while the descendants of European settlers, despite a prevalent discourse of attachment to the soil, frequently migrate (again) or switch economic pursuits if the opportunity arises. In a comparative perspective this latter attitude towards the land is a rather peculiar and maybe even "weird" constellation (see Henrich et al. 2010) driving a very particular expansive movement. Nevertheless, until now climate and natural resource models (see the HEP Human Potential as introduced by Klein et al. 2021) adopt this cultural stance. Biased by the European experience of the last few hundred years these models assume that people will move into a new habitat if it is available to them. Moreover, they assume that the hunting and gathering agents of the first move out of Africa will exhibit the same "rationality" as we find it realized in the colonial expansive movement.

A recurrent way of characterizing the difference between expansionist migratory movements of the (post)colonial era and hunter-gatherer contexts is an emphasis on the fission-fusion pattern found in hunter-gatherer mobility. The pattern has been described in detail elsewhere (for Africa see Barnard 1992, Blackburn 1996, Widlok et al. 2012, for a general overview see Marlowe 2005) and is depicted schematically in Figure 1. It is noteworthy that environmental features play a role in this pattern but not in a simple deterministic way. Barnard (1992), for instance, has highlighted that some foraging groups in southern Africa are in fission mode during the wet season (with water sources being abundant) and in fusion mode during the dry season (when people congregate at few permanent waterholes) but that the opposite also occurs in neighbouring groups: In extremely dry regions like the central Kalahari the environment only allows fusion in the wet season when there are sizable water sources at all, while the dry season requires groups to split up into smaller fractions. The pattern has also been observed for hunter-gatherers outside Africa, outside the tropics and subtropics. Here the fission-fusion pattern is even more marked given the considerable seasonal variation in resource availability (Wengrow and Graeber 2015). It has also been noted that despite the environmental dimension just sketched there are considerable social implications to this fission and fusion pattern as it may allow people to alternate between more hierarchical and centralized forms of group formation and more egalitarian and decentral ones (see Wengrow and Graeber 2015). Moreover, splitting and (re-)uniting occurs amongst foragers throughout the year and often also for non-ecological reasons, primarily for reasons to do with conflicts and conflict-resolution (Widlok 2016, 2017a, 2017b). As a whole the fission-fusion pattern explains why many hunter-gatherers can survive in very marginal environments and it is above all an explanation for the limited mobility which brings them back more or less in circles as they move between a number of possible sites within a region. Although hunter-gatherers of today, and of the recent past, would typically establish new abodes a little distance away from huts that were built in the past when returning after a year or two, their mobility is largely one of re-visiting places within a region and within a lifetime and not one of out-migration as found in the established frontier models.

However, the fission-fusion model has also received some criticism recently (see Bird et al. 2019) because not only is it commonly interpreted as an

Figure 1: The conventionally idealized fission-fusion pattern across three stages, persons group around waterholes, they disperse and re-aggregate in the following season.

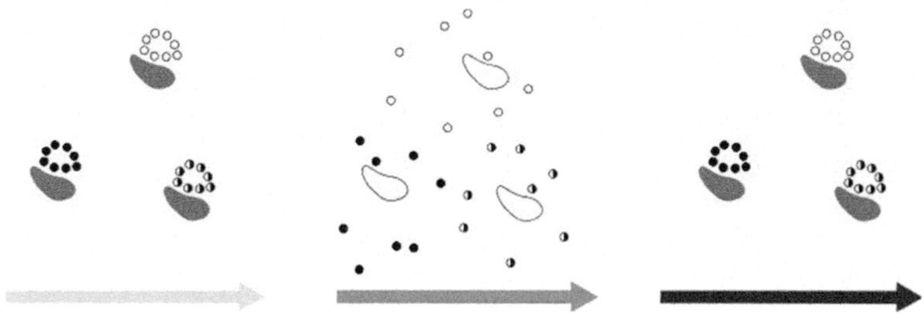

environmental deterministic mobility pattern but one that follows the changing local distribution of resources over time. In other words the fission-fusion model fits the image of relatively isolated small-scale societies that react to local environmental changes but which remain largely unaffected by larger scale networks. It also provides no prospect of how human populations could have broken out of this dynamic equilibrium pattern that they have inherited from non-human primates (see Dunbar et al. 2014, and as a critique Wengrow and Graeber 2015 and Graeber and Wengrow 2021: 279).

In our project we have investigated two ways of re-evaluating the fission-fusion model in terms of the relevance of larger scale networks. One is the embedding of the fission-fusion dynamic into that of (larger) developmental cycles (Widlok et al. 2012) and one is to revisit the model in the light of detailed ethnography. A key challenge was to account for any directional movement that would take foragers outside and beyond their home region. In the ethnographic record such directional migrations among foragers have only been observed in situations of colonial pressure, displacement and resettlement where groups were forced out of an area either by the colonialists themselves or by other African groups who had previously been pushed out of the area they were occupying (see Weig 2013). How can a model that successfully represents the resilience of land-based regional mobility that reproduces occupancy over time account for the fact that apparently in some cases there has been a transformation of that pattern into an "outward" movement, however slow and haphazard this may have been? Are the mechanisms that govern fission and fusion sufficient to account for a wider dispersal as well? How can we conceptualize such a dispersal that still preserves the hunter-gathers' ties to their environment, their perception of the environment as one that is "giving" (Bird-David 1990)? We have given a partial answer by connecting the fission-fusion model with the idea of "adaptive cycles" (Widlok et al. 2012). The adaptive cycle idea allows us to see how a circular, homeostatic fission-fusion movement may be turned into an upward or downward spiral through external effects such as environmental change that may impact the system differently in different phases of the developmental cycle and lead to some transformation within an overall reproduction. Here we report on another part of the answer which uses agent-based simulations for observing long term spatial effects of fission and fusion movements that are generated from within the social system, in particular through the social dynamics of searching for appropriate marriage partners.

The Changing Group Composition Model

In optimal forager models, which are adapted from animal behaviour studies, individual agents need someone, anyone, to marry (or mate) in order to reproduce. And they need someone with whom to economically cooperate. Again, this could be anyone. In real life situations of hunter-gatherers the situation is more complex. People are restricted to whom they can marry (to be discussed below); they are also restricted with whom they reside and collaborate. These restrictions channel various options to engage in social relations, facilitating some forms of living with one another and restraining others. Ethnographers usually emphasize that in comparison with many agriculturalists the hunter-gatherer mode of life allows for considerable flexibility in group composition and individuals make ample use of this flexibility as they "vote with their feet", making use of their freedom to move away. There is a degree of mutuality, collaboration, and support so that the social system of most ethnographically documented African hunter-gatherers is by and large geared against specific social obligations that would lock individuals in specific households and village communities of the type that characterize agricultural systems in Europe (see Sabean 1990).

Hunter-gatherer mobility is also embedded in social bonds that constrain but also enable their shifts in residence – bonds that are dispensable for explaining the behaviour of non-human animals for which models of OFT have first been developed. Kinship in the human settings is largely performative, which means that individuals can forge and select some links (and allow others to lapse) through their actions such as frequent visiting, assisting, gift-exchange and sharing. Marriage ties, too, can be dissolved fairly easily and most people have more than one partner in the course of their lives. But in contradistinction to non-human contexts, social ties such as marriage or siblinghood do inform the residential patterns of hunter-gatherers (Woodburn 1968). Individuals (re)adjust carefully to the social expectations but also to the requirements of particular situations. Even children from an early age onwards have some control of who they live with, many end up spending a great deal of time with their grandparents. Hunter-gatherer bride service arrangements involve staying with parents-in-law for a while but unlike the bride wealth payments common in agricultural systems this does not create lasting dependencies of the more junior on the more senior ones. Beyond the regularities of kinship, people in these societies also undergo great trouble to maintain particular friendships, for instance through gift-exchange partnerships across time and space (Wiessner 1982).

In other words, despite a large degree of flexibility, it does matter for people with whom they share a camp. For a long time this basic fact has been overlooked when theorizing about the fission-fusion pattern. All that counted was the overall number of residents in a local camp at any one point in time (see Figure 1). The numbers seemed to even out roughly in each season which makes the system appear to be locked in homeostasis. The ensuing pattern looks more stable and immune to transformations than it actually is. When we re-introduce the identity of individuals to the model, a very different picture emerges. In Figure 2 we have "specified" camp members through colour coding. Here, despite a stable growth and decline cycle, there is considerable change in the actual composition of these groups as only a few of the "original" people reconvene at a place in the next fission or fusion phase of the cycle. From the perspective of any one individual in such a fission and fusion system there are therefore considerable changes with every "seasonal" move. Even if the size of the local group returns to its seasonal "normal", the size of the group any one has lived with (and may return to live with in the future) has actually increased. Given the high flexibility and high permeability of local groups, individuals come to live together with a much higher number of diverse individuals over time than at any one point in time because the cards are shuffled again in each season. The net effects of this pattern is that the individual networks are reaching much further than the places that the person him- or herself has been visiting because former co-residents (at the same time potential co-residents to be) are spread across a larger area and individuals can activate the larger networks if necessary.

Figure 2: The refined fission-fusion pattern in the Changing Group Composition Model: persons around waterholes disperse and re-aggregate in different compositions across seasons.

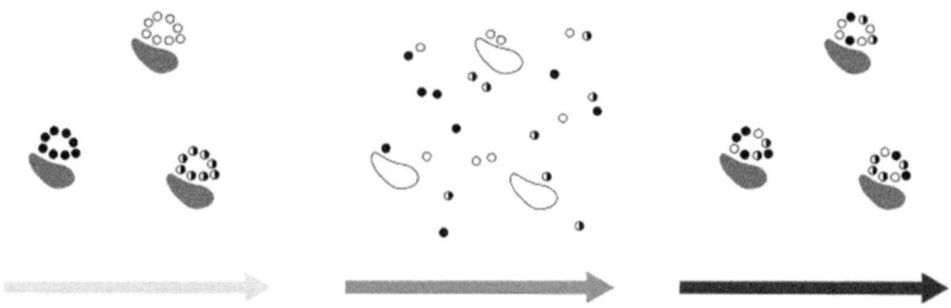

This more realistic type of modelling solves an issue that has been puzzling researchers for considerable time: Despite living in small groups, very small groups indeed at times of the year, the groups that people identify with and that they can rely on in terms of searching for partners (in marriage, in exchange, and in collaboration) is actually much larger. The local organization is not to be confused with the social organization (Bird et al. 2019: 98). The former comprises who is with whom at any given point in time while the latter includes the "expansive and virtual patterns in ties that comprise networks of social interaction" (Bird et al. 2019: 98). Those networks may be extensive networks of gift-exchange as in the San *hxaro* system (see Wiessner 1982), networks of ritual affiliation, or yet other links but the effect is the same: Even when only living together in camps of around 25 individuals, the number of people one is in direct or indirect contact with over a lifetime can be at around 1000, and an average adult may be in contact with well over 300 other adults and, for instance, their particular styles of working a tool (Bird et al. 2019: 97-8). The ensuing large networks are also used to transport items (material objects but also religious cults or mythical themes) across whole continents (see Bird et al. 2019). These links make the society also much more resilient to environmental, ecological, or demographic crises than a very small local group could ever be. What is more: The established idea that small residential groups are recruited out of homogeneous "small-scale societies" turns out to be a serious distortion in the light of evidence that small foraging groups more often than not consist of affines and not of close blood relatives or spouses (Bird et al. 2019: 102). As is the case with indigenous Australians, even when only very few individuals come together in foraging groups or hearth groups, it is not uncommon at all to have several dialects, ritual affiliations, and diverse links to many places in an open network beyond ethnic or linguistic groups represented (see Bird et al. 2019). Therefore, Bird et al. (2019) conclude that this makes "small" foraging groups in fact large and complex. It also makes them inherently "outward" looking, not only when looking for spouses or new patches to forage, as they are genuinely connected in a wide network. The implications are far reaching in that transgenerationally forager groups with little material accumulation can be said to accumulate the social capital of being connected. Moreover, we will have to give up the idea that large scale mobility of small groups due to their size will have to be channelled through external environmental factors. Given that foragers are aware of a much larger number of fellow humans and their environmental effects than what their local group size suggests, they also orient their movements accordingly. Bird et al. (2019) are able to show this with regard to burning practices (to enable better hunting of certain species). These practices shape the land at a large scale and feed back into the decisions in small foraging bands to move (or stay). In this context a piece of land that carries traces of being shaped by humans through burning (possibly across generations) can be more attractive than seemingly "free" land that has received less attention and is less well prepared for foraging activities. An adequate depiction of a "human existence potential" (HEP, see above) would need to include the effects of such large-scale transgenerational networks and land-uses on hunter-gatherer decision-making.

The other important feature that emerges from this enhanced model of changing compositions is that it lays the foundation for transformations in other ways. It lays the foundation of creating larger settlements, if the conditions are right, since the larger network that is already there in a dispersed fashion could in principle concentrate at a single place. An example for this are the ritual men initiation camps of San groups that were held in winter every few years and could encompass well over 200 persons (Lee 1979: 365). Other renowned examples are the Nambikwara with seasonal hilltop villages comprising several hundred people, and the larger seasonal settlements of the Inuit and the Great Plains Indians (see Wengrow and Graeber 2015, Graeber and Wengrow 2021). It also lays the foundation for an out-migration of sorts, if the conditions are right and even without any ideology of "looking for greener pastures elsewhere" or some assumed "Wanderlust". It would suffice to create such a movement out of an aggregated effect of people following the pattern that they have been following anyway, namely re-convening in newly composed groups under conditions that may create some sort of drift across local groups with every move. If individuals, for whatever reason, were choosing to live with people they had lived with before, except that they choose those who happened to be at the extreme end of a known and frequented area, this could easily have the cumulative effect of a drift-type movement – without any particular changes to the economy, the ideology or the socio-political make-up of the group being necessary. Add to the picture that in every generation, and from an early age on, there is considerable flexibility in choosing where to live (see above) and a transformation can ensue simply through reproducing the established and entrenched rules.

Thus, the pattern of changing the composition of local camps in a fission and fusion dynamic could lead to such a drifting "migration" without disrupting the established system and without assuming that hunter-gatherer adopted an expansionist mode of living or mind-set similar to that of agropastoral farmers or frontier settlers.

Adding marriage rules to the model

However, up to this point we have only shown that fission and fusion dynamics are compatible with transformations. But why should such transformations be likely to occur? We propose that in addition to environmental conditions, there are social reasons that may drive transformations. More specifically we have in mind the effect of marriage rules.

While we have no way of knowing which marriage rules exactly foragers of the past have followed, it is very unlikely that early Homo sapiens had no such rules at all because marriage rules are one of the few and best-researched human universals that do characterize human life across time and space (see Antweiler 2007: 7). We therefore have made these rules part and parcel of our agent-based modelling in which we simulate hunter-gatherer mobility and sociality (see Henn forthcoming). Marriage rules do differ culturally, not only with regard to what exactly the restrictions entail (e.g., whom one may marry and have sexual relations with) but also with regard to who the preferred marriage partners are. Whatever the exact rules are, the important point to realize is that the introduction of any rule immediately and considerably enlarges the demographic size a group (or a social network) needs to reproduce successfully.

One of the key features to distinguish human kinship systems is whether "cousins" (the children of the siblings of one's parents) are considered siblings and marriageable or not. At times only cross cousins (children of MB and FZ) are considered cousins while parallel cousins (children of MZ and FB) are not. In other systems all cousins are considered cousins and equally marriageable – or all of them may be considered siblings and therefore non-marriageable. With regard to marriage preferences again sometimes cross cousins are considered preferred marriage partners and sometimes it is parallel cousins, and sometimes marriage beyond all first cousins is considered desirable. Whatever the detailed rules that restrict the choice of potential marriage partners in one way or another, our point is that individuals always need a much larger group to choose from.

For individual persons, marriage rules always entail the need for a larger personal network: As marriage rules exclude certain others as possible spouses, individuals have to look for spouses further away in terms of kin relatedness and space. For populations, marriage rules entail that in order to be demographically viable – to be resilient against stochastic fluctuations – they need to be larger.

Among contemporary social groups there is a wide array of residential and marriage rules, and we cannot know for sure which of these cultural rules were applied and followed by the people who lived during the first human dispersal between source area and sink area on the way from Africa to Europe at various stages in the past. At the same time, as sketched above, human variation is limited in this domain allowing us to model several possible scenarios and outcomes with the help of agent-based simulations. Introducing such rules allows us to see the (unintentional) long-term effects of individual choices. Running a model that simulates the effects of marriage patterns over 500 years, Stephan Henn (Henn forthcoming) is able to show the cumulative effects of what may appear to be rather unspectacular and small differences in marriage rules in terms of population dynamics. As pointed out above there are marriage and incest prohibition rules in all human societies, but they vary in a fashion that can be modelled – and thereby to see long-term consequences for population dynamics, and potentially also for migration. Henn compared three different artificial societies a) one with a basic incest rule (no marriage to siblings and parents); b) one with more complex rules, namely no marriage with first cousins; and c) one prohibiting marriage with second cousins (see Figure 3). The three societies differ in the number of individuals they must minimally comprise if they want to have a fair chance of survival (i.e. 75% of the simulated societies survive 500 years) over many generations given stochastic fluctuations in fertility and mortality. Results show a minimum population size of 125 for societies with first-cousin prohibition and a minimum population size of almost 200 individuals for second-cousin prohibition. A near to 100% survival rate is only achieved at even larger numbers. It is worthwhile remembering that this is the survival of a group under good and stable ecological conditions since no ecological crises have yet been factored in. These population sizes are far beyond the number of forager resident groups that were reported in the ethnography of Africa and beyond, which number only about 25 (Hamilton et al. 2001). Thus, the simulation of population dynamics, enhanced by marriage rules, provides sufficient reason to assume that residential groups cannot be self-sufficient but will always have to be part of a larger network. We can continue along these lines by enhancing the simulation further through not only including marriage rules but also the post-marriage resident rules that are usually an integral part of marriage regulations.

Figure 3: Minimal population sizes simulated with three different basic marriage prohibitions in place.

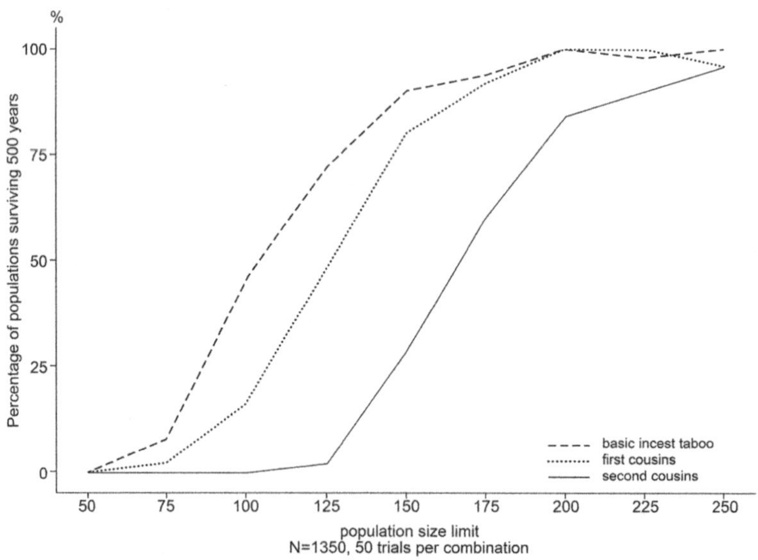

If we add mobility and residency to the simulation by means of a patrilocal post-marriage rule and if we further assume that the rule was followed without exception it only takes a few generations before people become concentrated at a few places (see Henn forthcoming). This socially motivated mobility entails a spatial distribution that deviates from an ideal free distribution – one that only takes (non-social) resources into account. The simulation suggests that in this case, different sites for habitation would be sought when resources become depleted, and some resource-rich places to live in between major settlement sites could remain unoccupied. These abandoned places could later become inviting for others who are switching residence after getting married. Thus, whatever marriage rule exists in such a small-scale environment, we can safely assume that it would inevitably lead to considerable mobility. The model allows us to go one step further: Given that a good chance for group survival only exists if there is a larger network around a residential group that is big enough for individuals to find a marriage partner, there is a major incentive for maintaining regional networks beyond individual residential groups. Going beyond the regular roaming area for searching for a partner is a necessary strategy in such contexts and settling elsewhere with that partner and the offspring becomes an option. But again, no break of existing mobility rules is required as people could simply use the flexibility and permeability that the fission and fusion system already provides. Personal network ties provide individuals and families with opportunities to link up to different residential groups in the region, equally small and equally flexible like the ones they originate from. There is no need to assume climate crises or other ecological problems that force people to migrate. Henn's simulation suggests that finding an appropriate marriage partner is reason enough for a considerable amount of mobility. The result is not that individuals undertake intercontinental migration, but that there is a regular incentive for a good proportion of the group in each generation (!) to move beyond the area in which they were brought up. Over a few generations mobility across regions can emerge from a pattern of localized mobility. While there are different estimations as to how quickly migration out of Africa took place (see Litt et al. 2021), all estimations suggest that it took many more generations than what our simulation would suggest as minimally necessary for such an emergent directional mobility. Over time any marriage rule will lead to people leaving the group to find partners elsewhere: they will move even if they live in the most splendid natural environment and favourable climate conditions, and they may move even if the place they move to is less favourable in terms of natural resources. Finding an appropriate partner may easily trump having a more varied diet on your plate. The internal social drivers that we propose here are sufficient to explain a gradual "move" out of Africa. There is no need to assume environmental disasters or an ideology of Wanderlust or of exploration and exploitation as a driving force. Post-nuptial (post-marital) residence rules add to the mobility induced by marriages. Again there is at this stage no way of telling what particular combination of residence rules that are found today was applied by any particular hunter-gatherer group of the past. They could have been matrilocal (near the maternal family), patrilocal (near the paternal family), uxorilocal or virilocal (near the wife's or the husband's home place) or neolocal (a residence independent of descent). But again, it is likely that some such rule was applied, and we can model what happens when various rules are being followed. For instance, the more a group sticks to a patrilocal post-marital residence rule, the less people become dispersed, and ultimately they all end up in one single place. The more a group extends the bride service rule, the more people become dispersed, and extended exchange between residential groups will ensue.

We do not deny that external drivers of mobility can play a role, but we feel the need to emphasize that endogenous dynamics may suffice. Humans are very much social beings whose interdependence cannot be factored out without missing a central piece of what makes us human.

Conclusions: Upscaling as a forager technique

What we have seen so far is that firstly, that fission-fusion patterns can be read and transformed to be non-homeostatic rather than be always confined and static. We have also seen, secondly, that with the cultural establishment of marriage rules there is a systematic incentive for agents to seek opportunities for expanding their networks. All this, it could be argued, merely shows that a major shift from small-scale to larger-scale societies (in terms of their mobility pattern) is possible within the hunter-gatherer spectrum. However, that would still make it a considerable cultural shift where quantity changes in the scale of social networks, if you like, lead to quality changes in terms of social relationships and sociality. Crudely summarizing Nurit Bird-David (2017) on this point, her argument is that there is a major cultural shift between the plurirelational persons in small-scale forager settings and the standardized mereological individuals (based on generic part-whole relationships) in larger settings, a shift that prohibits upscaling our terms and models. Or, to put it differently, modellers may be tempted to upscale features of forager society (including mobility) where the ethnography suggests qualitative thresholds below which relationships are categorically different and therefore resist upscaling. The argument is not only that scale influences social form (see Barth 1978) but more radically that small-scale may intrinsically inhibit "upscalability" and the practice of scaling itself. The issue is whether hunter-gatherers are engulfed in this small-scale world or whether their world may be said to always include the local group as well as the larger network at the same time. The larger structures need not manifest themselves as large gatherings of people, but they can still be evident and effective as in the virtual communities evoked in the firelight talks described by Wiessner (2014).

Where do we go from here in the light of what we have presented in this contribution? Attempts to simulate forager mobility in a way that is concurrent with present-day ethnography does not stop at marriage rules and residence rules. It is one of the typical features of such models (and of simulating artificial societies more generally, see Epstein and Axtell 2011) from the "bottom up" that they can incorporate ever new features of that ethnography which are not visible in the archaeological record. So far our simulations have placed foragers in an environment that is equipped with renewable resources but which is otherwise "empty". But hunter-gatherer specialists have noted for some time that such an assumption of an "empty" environment is likely to be a misrepresentation (see Marshall-Thomas 2006, Widlok 2019). Hunter-gatherers see other species not only as resources but as partners with whom they engage with. One could go as far as saying that what we have presented here is but one modality of "finding a partner", i.e. in this case a marriage partner. But a more complete picture would have to find ways to also include the search for other, non-human partners – which is different from treating plants and animals as passive and unrelated "resources". It has been pointed out that pre-colonial movements in Africa were not predicated on the European illusion (or fantasy) cultivated over centuries that people were moving into unoccupied territory, empty space (see Kopytoff 1987). This is even more so the case for the African foragers who are very likely not to separate the human domain from that of other species. There is ample evidence across the board that hunter-gatherers include other species into "the social". They not only entertain special relationships with many other species (see Widlok 2019 on the relation between San and lions) but more generally they consider non-humans to be part of the social world, of being legitimate occupants of land and of being social partners (see Sahlins 2017). Models that rely on calculations of the "Best Potential Path" and "Human Existence Potential" (see Klein 2021) are blind to any obstacles that are not based on physiography or climate, disregarding spirits and other species: Where foragers will see other species (not only game species) or non-human beings to either disable or enable their moves, path, routes and their existence at large, these models simply draw a line from A to B superimposing their own cultural ontology which is very selective in recognizing what can constitute an obstacle or a path. Even in an apparently underpopulated environment, the number of social relations that matter is in any case larger than the number of humans in a group, amplified by the fact that humans have also got relations with other beings in that environment. Whether this makes forager societies inherently more hierarchical (as Sahlins 2017 claimed) or not, it makes them certainly more populous.

However, future research on scale would need not only to rethink the number and kind of agents that need to be taken account of. Apart from quantity it would also need to reconsider the quality of relationships between these agents. As mentioned above, Bird-David (2017) suggests that only in large-scale "mereological" societies people recognize "types" of humans (or other beings) where small-scale foragers see plurirelational beings. But again the question is how the former could have developed out of the latter.

Along the lines of the argument that we have developed above we would hypothesize that hunter-gatherer social systems also provide prototypical examples of how generalizable relations emerge out of particularist relations. One of the features that Henn's simulation will eventually include is a way of modelling the effect of "name-sakes": While European philosophers and linguists tend to draw a sharp line between personal names and generic types, the naming system of the San, for instance, brings together both. Since there is only a limited number of personal names available that get "re-cycled" over generations, every person is both, the plurirelational self that Bird-David depicts but also, at least latently, a representative of a named category. As Henn (forthcoming) is able to show in his simulations, this, too has an immediate long-term effects in terms of mobility, social aggregation and dispersal because name-sakes shape the preferences that influence individual decision-making and mobility. Having the same name as an older San identifies a younger person with the older one and their kin relations, overriding all genealogical relationships of the younger one but those of the nuclear family (Lee 1986). Naming-relationships are an important attractor for producing the "colourful" fission-fusion pattern that we have outlined above. As with the San, Name-sake relationships influence marriage options. Since, for instance, the sister of my name-sake becomes my sister, she becomes non-eligible as a marriage partner. Personal naming systems provide the cognitive bridge towards classificatory systems with avoidance rules of the kind that we have outlined above. They allow a "plurirelational" system to move towards a "mereological" system. It remains to be seen whether this shift correlates with the shift from regional to outward mobility as we have discussed above. What it hints at, though, is that despite living in small numbers at any one time, hunter-gatherers themselves make use of upscaling in terms of "techniques of the intellect" (see Levinson 2020). They recognize people (and including non-humans) in their environment as members of generalized groups, as well as in terms of kinship relations, and this entails extensive mobility practices.

Acknowledgements

The research on which this chapter is based was funded by the Deutsche Forschungsgemeinschaft through the CRC 806 "Our way to Europe". Ralf Vogelsang and Karin Kindermann provided helpful comments to an earlier version of this paper which is based in large parts on Widlok (2021). The illustrations are based on the PhD work by Stephan Henn (forthcoming) and were drawn by him and with the assistance of Lutz Hermsdorf-Knauth. Special thanks go to Werner Schuck who greatly facilitated the smooth operation of the project throughout the funding periods. If there was ever any further confirmation needed about the importance of social relations, pragmatics and networks in major collective operations such as "Our way to Europe", Werner Schuck's ingenious support would prove the point.

References

Antweiler, C. 2007. *Was ist den Menschen gemeinsam? Über Kultur und Kulturen*. Darmstadt: Wissenschaftliche Buchgesellschaft.

Barnard, A. 1992. *Hunters and Herders of Southern Africa: A Comparative Ethnography of the Khoisan Peoples*. Cambridge: Cambridge Univ. Press.

Barth, F, (ed.) 1978. *Scale and social organization*. Oslo: Universitetsforl.

Bird, D. and Bliege Bird, R., Codding, B., Zeanah, D. 2019. Variability in the organization and size of hunter-gatherer groups: Foragers do not live in small-scale societies. *Journal of Human Evolution* 131: 96-108.

Bird-David, N. 2017. Before nation: Scale-binding anthropology and foragers worlds. *Current Anthropology* 58/2: 209-226.

———.1990. The Giving Environment: Another Perspective on the Economic System of Gatherer-Hunters. *Current Anthropology* 31/2: 189-96.

Blackburn, Roderic H. 1996. "Fission, Fusion, and Foragers in East Africa: Micro- and Macroprocesses of Diversity and Integration among Okiek Groups." In: S. Kent (ed.), *Cultural Diversity among Twentieth-Century Foragers: An Afri-can Perspective*, Cambridge: Cambridge University Press, pp. 188-212.

Brody, H. 2001. *The other side of Eden: Hunter-gatherers, farmers and the shaping of the world*. London: Faber.

Dieckmann, U. (ed.) 2021. *Mapping the Unmappable? Cartographic Explorations with Indigenous Peoples in Africa*. Bielefeld: transcript.

Dunbar, R., et al. 2014. "The road to modern humans: Time budgets, fission-fusion sociality, kinship and the division of labour in hominin evolution." In: R. Dunbar et al. (eds.), *Lucy to Language*, Oxford: Oxford University Press, pp. 333-55.

Eggert, M. 1998. Archäologie und Analogie: Bemerkungen zu einer Wissenschaft vom Fremden. *Mitteilungen der Anthropologischen Gesellschaft in Wien* 128: 107-124.

Epstein, J. and R. Axtell. 2011. *Growing artificial societies: Social science from the bottom up*. Cambridge: MIT Press.

Gladwin, C. 1989. *Ethnographic decision tree modeling*, vol. 19 of Qualitative Research Methods Series. Newbury Park: Sage.

Graeber, D. and D. Wengrow 2021. *The dawn of everything. A new history of humanity*. New York: Farrar, Straus and Giroux.

Hamilton, M., B. Milne, R. Walker, O. Burger, and J. Brown. 2007. The complex structure of hunter–gatherer social networks, *Proceedings of the Royal Society B: Biological Sciences* 274: 2195-2203.

Henn, S.. Forthcoming. *Placing decisions in context: the emergence of hunter gatherer social networks*. PhD dissertation, University of Cologne.

Henrich, J. and S. Heine, A. Norenzayan. 2010. The weirdest people in the world? *Behavioral and Brain Sciences* 33: 61-135.

Kelly, R. 1995. *The foraging spectrum: Diversity in hunter-gatherer lifeways*. Washington: Smithonian Inst. Press.

Kent, S. (ed.) 1996. *Cultural diversity among twentieth-century foragers: an African perspective*. Cambridge: Cambridge Univ. Press.

Klein, K. & Ch. Wegener, I. Schmidt, M. Rostami, P. Ludwig, S. Ulbrich, J. Richter, G. Weniger, Y. Shao. 2021. Human existence potential in Europe during the Last Glacial Maximum. *Quaternary International*, 581-582: 7-27.

Kopytoff, I. 1987. The internal African frontier: the making of African political culture. In: I. Kopytoff (ed.) *The African frontier: the reproduction of traditional African societies*. Bloomington: Indiana University Press, pp. 3-84.

Lee, R. B. 1979. *The!Kung San: Men, women, and work in a foraging society*. New York: Cambridge University Press.

———. 1986. !Kung kin terms, the name relationship, and the process of discovery. In: M. Biesele, P. Wiessner, R. B. Lee, & R. Gordon (eds.), *The Past and future of!Kung ethnography: Critical reflections and symbolic perspectives: Essays in honour of Lorna Marshall*. Hamburg: Buske, pp. 77-102.

Levinson, S. 2020. *On 'Technologies of the Intellect'*. Goody Lecture 2020: Max Planck Institute for Social Anthropology.

Litt, Th., J. Richter; F. Schäbitz (eds.). 2021. *The Journey of Modern Humans from Africa to Europe*. Stuttgart: Schweizerbart Science Publishers.

Marlowe, F. 2005. Hunter-Gatherers and Human Evolution. *Evolutionary Anthropology* 14/2: 54-67.

Marshall-Thomas, E. 2006. The Lion/Bushman relationship in Nyae Naye in the 1950s. In: J. Solvay (ed.), *The politics of egalitarianism: Theory and Practice*. Oxford: Berghahn Books, pp. 119-130.

Polanyi, K. 1944. *The great transformation*. New York: Rinehart.

Sabean, D. 1990. *Property, production, and family in Neckarhausen, 1700-1870*. Cambridge: Cambridge University Press.

Sahlins, M. 2017. The original political society. *Hau: Journal of Ethnographic Theory* 7/2: 91-128.

Turner, R. 1893. *The Significance of the Frontier in American History*. In: R. Billington (ed.), Frontier and Section: Selected Essays. Englewood Cliffs, N.J: Prentice-Hall, pp. 28-36

Weig, D. 2013. *Motility and relational mobility of the Baka in North-Eastern Gabon*. PhD dissertation University of Cologne, available online. Köln: Universitäts- und Stadtbibliothek.

Wengrow, D. and D. Graeber. 2015. Farewell to the 'childhood of man': ritual, seasonality, and the origins of inequality. *Journal of the Royal Anthropological Institute* 21/3: 597-619.

Widlok, Th. 2021. *Hunter-gatherer mobility*. In: Th. Litt, J. Richter; F. Schäbitz (eds.), The Journey of Modern Humans from Africa to Europe. Stuttgart: Schweizerbart Science Publishers, pp. 41-51.

———. 2019. Human-lion relations. In: Th. Breyer and Th. Widlok (eds.), *The situationality of human-animal relations*. Bielefeld: transcript, pp. 203-219.

———. 2017a. "The decision to move: being mobile and being rational in comparative anthropological perspective." In: P. Meusburger, B. Werlen, and L. Suarsana (eds.), *Knowledge and Action*. Cham: Springer Open, pp. 253-65.
———. 2017b. "Die eigene und die fremde Mobilität." In: M. Wunsch and G.-C. Weniger (eds.), *2 Millionen Jahre Migration: Wie(so) wir darüber forschen und reden*. Mettmann: Neanderthal Museum, pp. 64-71.
———. 2016. Moving between camps: towards an integrative approach to forager mobility. *Hunter Gatherer Research* 1/4: 473-494.
———. 2000. "On the other side of the internal African frontier: relations between Herero and 'Bushmen'". In: M. Bollig and J.-B. Gewald (eds.), *People, Cattle and Land*. Köln: Köppe, pp. 497-522.
Widlok, Th. and A. Aufgebauer, M. Bradtmöller, R. Dikau, Th. Hoffmann, I. Kretschmer, K. Panagiotopoulos, A. Pastoors, R. Peters, F. Schäbitz, M. Schlummer, M. Solich, B. Wagner, G.-C. Weniger, A. Zimmermann. 2012. Towards a theoretical framework for analysing integrated socio- environmental systems. *Quaternary International* 274: 259-272.
Widlok, Th. and K. Eghbal-Azar. 2012. Potentials and limitations of mobile eye tracking in visitor studies: evidence from field research at two museum exhibitions in Germany. *Social Science Computer Review* 31/1: 103-118.
Winterhalder, B. and Smith, E.A. 2000. Analyzing adaptive strategies: Human behavioral ecology at twenty-five. *Evolutionary Anthropology*, 9: 51-72.
Wiessner, P. 2014. Embers of society: Firelight talk among the Ju/'hoansi Bushmen. *Proceedings of the National Academy of Sciences*, 111/39: 14027-14035.
———. 1982. "Risk, reciprocity and social influences on !Kung San economies." In: E. Leacock and R. Lee (eds.), *Politics and history in band societies*, Cambridge: Cambridge Univ. Press, pp. 61-84.
Woodburn, J. 1968. "Stability and Flexibility in Hadza Residential Groupings", In: R. Lee and R. de Vore (eds.), *Man the hunter*. Chicago: Aldine, pp. 103-110.

Comment by Robert L. Kelly

Widlok and Henn seek to help answer the question: How did a migration to Europe emerge from societies that were mobile but not migratory in the narrow recent sense?

Their answer rejects approaches to mobility that privilege environment and subsistence, and consequently optimal foraging theory. They claim that ethnography "has taught us to mistrust models of mobility that are purely resource-driven" even though I would claim that ethnography has actually provided abundant support for such models (Kelly 2013) – at a certain scale of analysis. They also cancel foraging models because they "emerged in a particular cultural situation, enlightenment Europe, which was predicated economically and politically on an expansionist, imperial and growth-oriented culture." A migratory wave into Europe is likewise rejected as merely reflecting a Eurocentric bias toward a culture that was "ever seeking greener pastures". Given this reasoning, should we cancel everything the enlightenment achieved? Surely not. And is it possible that the causal arrows go the other way: perhaps the imperialistic, expansionistic, growth-oriented culture of Enlightenment Europe was a product of those tendencies and capacities captured by optimal foraging models, an ever-present desire and intent to maximize advantage.

In fact, there is good justification for believing that optimal foraging models capture something fundamental about humanity that led our species to colonize nearly the entire globe while living as hunter-gatherers.

The biological capacity for culture (whatever it is) must have been driven by natural selection: an organism with such a capacity outcompeted conspecifics without it. Why? At its heart, culture is a model of how life is supposed to operate. But life never operates the way we want it to, there is always a disconnect, sometimes greater than at others, between how we think life should be lived and how life is actually lived. A conscious organism will likely strive to make life-as-it-is-lived closer to life-as-it-should-be-lived. Such an organism will have a selective advantage because it will always be consciously seeking "greener pastures": new knowledge, such as medicinal uses of plants, and new land (and, in the Paleolithic case, whose resources are untouched). Therefore, the capacity for culture might imply that restless exploration might very well be an adaptive feature of the human psyche. A feature, incidentally, that would be stymied among modern foragers living within boundaries imposed by the nation-state in the colonial and post-colonial world.

The real question is whether the models, whatever their origin, offer some insight and understanding of hunter-gatherer mobility. Ethnographic data suggest they do (Kelly 2013); studies using the Ideal Free Distribution, for example, show that people do move into a new habitat if they perceive the benefit to be worth the cost of doing so, and climate-induced environmental change coupled with population growth play roles in that process.

The purpose here is to use an agent-based model to see what effect marriage practices have on forager outward migration (because it is clear foragers did migrate). In effect, this chapter sees migration as an emergent phenomenon, something that arose unintentionally from another set of intentional behaviors.

Henn's efforts show that the fission-fusion pattern results in a mixing of people, such that, through the relationships developed, an individual's social network reaches to more geography than any individual has personally reached. And these linkages "make society also much more resilient to environmental, ecological or demographic crises than a very small local group could ever be." I agree – in fact, this is something anthropology has suspected since the 1966 Man the Hunter conference.

More to the point, the fission-fusion model that results in recombined local groups, comprised largely of affines, "lays the foundation for an out-migration of sorts, *if the conditions are right* ..." (Emphasis added). Those conditions would seem to be crucial, but they are not explicated here. Instead, "If individuals, for whatever reason, were choosing to live with people they had lived with before, except that they choose those who happened to be at the extreme end of a known and frequented area, this could easily have the cumulative effect of a drift-type movement." So, migration would be random, in both direction and timing. This proposal could also be simulated and compared to empirical cases to see if it could (for example) account for the (rapid) rate at which the New World was apparently colonized (even taking different colonization times into account). My guess is the answer would be no.

The marriage model also argues that the simple operation of a system in which people seek mates within a system of any rules (patrilocal, matrilocal ...), combined with a fission-fusion model, will eventually produce mobility. The agent-based model shows that people will become geographically concentrated and at some point unoccupied places will "become inviting." There is a lot wrapped up in that claim, a lot that probably has to do with the effects on foraging return rates of slow population growth coupled with climate-induced environmental change (something for which the Ideal Free Distribution model is suited). There are two scales that need to be merged into a single model here to evaluate how cultural practices could result in unintentional migration. That model should respect the effects of those cultural practices, such as marriage, and one that respects the material needs of a dynamic population responding to a dynamic foraging environment. I shall look forward to that.

References

Kelly, Robert L. 2013. *The Lifeways of Hunter-Gatherers*. Cambridge: Cambridge University Press.

Scales of interaction
Quantity and quality of encounters amongst northern foragers

Charlotte Damm

Introduction

Most past and recent foraging societies are described as small-scale, a term that often refers to both the size of residential groups and of any larger unit of socio-cultural interaction. Recently, this perspective was challenged by a study suggesting that while residential units may be small, all individuals are part of large-scale social networks, and basic units such as households or residential groups are not nested in a stratified set of socio-cultural groups (Bird et al. 2019). On the other hand, it has also been argued that scholars have neglected to understand the qualities and intimacies of small foraging groups (Bird-David 2017a, 2017b). Here I argue that a narrow focus on quantifying populations at any demographic scale diverts our attention from the dynamics and complexity of interaction within and beyond assumed entities or networks. Interaction operates at different geographical and temporal scales varying from local level and daily occurrences to distant and irregular interactions, and the quality and intensity of the interaction varies accordingly. In addition, even in numerically small communities the patterns of interaction may be very diverse, encompassing a variety of overlapping networks depending on gender, age, tasks, kinship and personal preferences.

While anthropology has often provided archaeology with models for social structures of foraging communities, archaeology is well positioned to address issues of scales of interaction. The input from archaeology is particularly relevant because much of archaeological data comes from foraging societies that were not encapsulated by farming communities nor were impacted by modern colonization. Using archaeological discussions concerning communities of practice combined with recent results from an ongoing research project in northern Norway as an example, I will argue that considering the flexibility and dynamics of foraging groups brings us beyond a fixed scale to instead reflect upon networks and interactions at a multitude of scales, varying from residential units to regional and long-distance contacts.

From spatial models to human interaction

The portrayal of hunter-gatherer communities as consisting of bounded local bands or residential units integrated in a larger entity representing a distinct socio-cultural unit is largely obsolete in contemporary anthropological research. Such traditional models appear more deeply ingrained in archaeology (Burke 2021). Explicit or implicit assumptions of a nested spatial organisation associated with socio-economic organisation have prevailed longer. However, few present it as constituting a socio-political organisation and a direct translation into cultural units is abandoned. Inspiration frequently comes from ethnographies. Burch, for example, writes of compound families within socio-political nations in his study of the Iñupiaq (Burch 2006) and the geographer Collignon presents the organisation of the Inuinnait as consisting of residential groups, with those exploiting the same territory forming a distinct and named community (Collignon 2006: 21). In her study of the mid-Holocene socio-spatial organisation of northern Sweden, Lundberg (1997) refers to June Helm's analysis of the 20th century Dene (Helm 1965) and employs the terms local and regional bands. Investigating socio-cultural divisions amongst foraging groups in Neolithic western Norway Bergsvik employs both ethnographic studies and research into ethnicity to suggest the existence of local territorial groups and identities (Bergsvik 2006).

Unfortunately, as many archaeological studies focus on spatial organisation in the landscape, they often neglect consideration of social and demographic flexibility in the composition of residential units or relocation between regions. The spatio-demographic organisation with a minimal band of 25-30 individuals and several such bands nested within a regional band exemplified in Whallon's heuristic model (Whallon 2006, fig.4) is therefore still familiar to many archaeologists, if not necessarily agreed upon. While Whallon's point is precisely that people do interact across units at all levels, the image of bounded, non-overlapping and nested units has not been fully replaced. Archaeology needs to consider the possibility of flexibility in choices regarding socio-demographic practices including residence and mobility patterns.

Communities of practice and learning networks

To emphasise the flexibility amongst past foragers, I propose that we turn to the insights from studies of transmission of technology and communities of practice. Individuals in small-scale societies would engage in a range of different activities and the sharing of related knowledge and skills. For the purpose of this chapter, a community of practice (Lave and Wenger 1991; Wenger 1998) is understood as a group of individuals who perform a specific task, sharing knowledge and techniques through a set of particular practices, even if dispersed between different sites. Such practices include manufacture of tools, equipment, dwellings and clothing, and hunting and fishing techniques (e.g. Jordan 2015), but also socio-cultural practices such as rituals, narratives and performances. We can then envision a wide range of communities of practice manufacturing and using specific items.

The practices are transmitted to others, typically the next generation, with varying accuracy, but generally following the same technological principles. Deviances could be due to local innovation or contact with persons from other communities of practice with alternative techniques, although the competency of the apprentice is also relevant. As emphasized by Gosselain (2000, 2008, 2010) some elements of a finished product may be easy to copy; for example, the shape and decoration of ceramic pots that perhaps need only be observed to be replicated. Other elements require more detailed information (e.g. mixing the clay and firing the pot), and therefore must be learned from accomplished individuals over a period of training or replicated using alternative techniques. When studying the spread of material items and practices, we should distinguish between such easily copied or transmitted elements which indicate interaction, but say little of the intensity of it, and elements that would have demanded a prolonged and more intimate learning period for transmission of skills and knowledge (Damm 2012a).

We could also describe such communities of practice as learning networks within which practices are reproduced, consolidated, and transmitted through repetition of technological actions. These networks would overlap only partially, as separate activities would probably have involved a different set of individuals. In small foraging groups with limited specialisation, each individual performs a variety of tasks. However, the network may consist of a slightly different group of individuals for each task. It follows that an individual interacts with different individuals for different purposes at different times (Damm 2012b). Much as a person may have multiple identities and be multirelational, s/he also participates in multiple, partly overlapping practice networks. In this way, a person may transmit a particular skill to some persons, and other skills to others. Skills and knowledge are not necessarily shared equally with all persons one encounters or lives with, but predominantly with people who engage in similar activities and share information. For example, Bird-David reports that while names of the most common plants were shared by all in a particular Nayaka hamlet, names of many other plants mentioned by some in the community were unrecognized by others. Individuals appeared to name plants differently, probably a result of the Nayaka practice of foraging "separately together", often in small groups consisting of a couple and their children (Bird-David 2017b: 128, 146). The transmissions of plant names would then be rather limited and in this case probably predominantly transferred from parent to child.

Gardner (2019) stresses that, as most of us, contemporary foragers weigh the information they receive depending on the reliability of the person sharing the knowledge. Aspects to be considered include the skills of the person providing the information, whether the information results from first-hand experience, and how trustworthy the person relaying the information might be. Evaluation of these aspects is easier with socially close individuals than socially distant persons. An ethnographic example from the Nayaka demonstrates the gradual integration of an in-married partner, requiring time for all involved to engage with each other, showing that multirelational ties and closeness evolves over time (Bird-David 2017b: 189). The more time spent together, and the closer individuals collaborate with one another, the bigger the impact they are likely to have on each other. While the recovery of such details may at first appear unattainable for archaeology, the combined use of insights from cultural transmission studies, communities of practice and learning networks will in some instances allow us to uncover the extension of close communication and transmission (Apel 2001; Hallgren 2008; Jordan 2015).

Spatial Demography of a northern maritime forager community: background

To discuss the extent of interaction within a prehistoric foraging group, it is necessary to explore both the quantity and quality of interaction at several scales, from individual households to long-distance contacts. Such a multi-scalar study requires the use of a range of different approaches, adjusted to the available data to infer quantitative and qualitative aspects of group sizes and interaction. Here I do not focus on the methodologies, but these cover traditional archaeological investigations of sites and artefacts, statistical analyses and comparisons with anthropological studies.

I will focus on the mid-Holocene maritime foragers of northernmost Norway, with an emphasis on western Finnmark (Damm et al. 2020). The geography deviates significantly from the inland contexts put forward by Bird-David (2017b) and Bird et al. (2019). Prehistoric foragers of the area had, since initial human colonisation in the early Holocene, inhabited coastal areas and had a distinctive maritime subsistence base, involving a strong reliance on boats and limited exploitation of terrestrial landscapes and resources. Northern coastal foragers often display elaborate technologies, not least associated with boats and clothing (e.g. Kelly 2013).

The mid-Holocene period c. 5500-2500 cal BC provides a particularly rich archaeological record for the region. Due to a slow rate of sediment accumulation and limited modern infrastructure, dwelling remains in the form of tent rings, cleared floors and house-pits are still visible on the surface. New analyses, based on a Summed Probability Distribution (SPD) of radiocarbon dates, indicate a relative population increase in northern Norway from about 6000 BC with a peak between 4500-3500 BC (Jørgensen 2020). While such models should be interpreted with caution, the initial population increase coincides with the onset of a more stable and predictable mid-Holocene climate, providing a likely ecological basis for population increase. From c. 5000 BC onwards, there is a marked increase in visible dwelling remains, demonstrating investment in more substantial structures in carefully selected locations (Damm et al. 2021). The distribution of distinctive artefact types indicates increased regionalisation. Overall, we assume that the mid-Holocene saw the development of a semi-sedentary settlement organisation, with prolonged seasonal stays at favourable locations. The many visible dwelling remains allow us to reconstruct the spatial organisation of settlements, albeit the focus on substantial dwellings emphasize occupation of some duration and exclude occupation with more ephemeral shelters (light tents, overturned boats, etc) and shorter stays.

Geographic and economic setting

Western Finnmark lies in the northernmost part of Norway and is characterised by long fjords and numerous sounds with a rugged and rocky coastline. The outer coast is sparsely vegetated with mainly shrubs and some birch, while the vegetation further into the fjords is dominated by birch, although pine was more plentiful during the mid-Holocene (Sjögren & Damm 2019).

Osteological data from excavations (Engelstad 1984; Hodgetts 2010) and evaluation of locally available resources (Damm et al. 2021) indicate an abundance of cod, seal and seabirds as the most frequent subsistence species, although reindeer, elk and fur animals were also exploited. It is noteworthy that, in contrast to other northern circumpolar areas, resources are plentiful on a year-round basis, with no marked lean seasons. A detailed study of one compact region on southwestern Sørøya in western Finnmark shows that sites were located within local seascapes such as bays, inlets or narrow sounds with easy access to resources, giving priority to fish, seal and terrestrial resources in that order of quantity, predictability and distance to foraging locations (Damm et al. 2022). Judging from the seasonal availability of resources it would have been possible to obtain all annually required subsistence resources within this small study area measuring c. 500 km^2 (or 650 km^2 if fjords and adjacent open sea are included). A few terrestrial resources, such as reindeer hides and antler in larger quantities may have been necessary to obtain from further afield. However, it was likely not foraging needs, but rather demographic, social, political and non-subsistence economic factors that motivated mobility out of, and away from, this area.

Spatial organisation, flexibility and mobility

To explore the demography and patterns of interaction amongst these northern foragers we need to investigate the spatial organisation at several scales including the size of residential units, and the geographical extent of regional and long-distance networks. Having indicated scales of spatial organisation (small residential units in a local seascape, regional networks covering several hundreds of km and long-distance contact at a scale above that) we must consider the extent of flexibility and mobility at several scales to understand the dynamic of interactions.

Residential units, seascapes and seascape groups

The number of dwelling remains at each site can vary from one or two up to 100, with 10 to 20 being most common. Early interpretations considered the sites as representing small villages, whereas research in the 1980s and early 1990s argued for an accumulation of successive dwellings (Helskog 1984), with possible regional anomalies and larger communities mainly after 2500 BC (Schanche 1995). The latest research confirms that not all dwellings were contemporary. Analyses employing Bayesian statistics on the radiocarbon dates from western Finnmark suggest that within a timeframe of 200 years, one to six dwellings at each site were inhabited, but the chronological resolution cannot answer how many of these were in use at the same time (Vollan forthcoming). Recent detailed surveys indicate that dwellings were often organised in small groups of one to four dwellings within sites. This suggests that generally only a small number of dwellings, likely two to four, were inhabited simultaneously. The interior floor area in the mid-Holocene period varied between 8-20 m^2 with an average of about 12-13 m^2, and often had a central stone-lined fireplace. With households possibly varying between four and ten persons, the residential unit size at such a site may have then ranged between 10-30 persons, with the average possibly on the low side of the magical number of 25 (Kelly 2013).

Sites were not evenly distributed in the landscape with long stretches of coastline uninhabitable due to steep cliffs. The geography of northern Norway and locational preferences structured habitation into distinct local seascapes with local resource exploitation (Damm et al. 2021). In such seascapes, spanning 1-2 km of coastline, there would be several possible habitation sites (Figure 1). In western Finnmark, inhabitable seascapes were concentrated in clusters 20-40 km apart with the area in-between often characterised by rough waters and limited landing sites (see also Figure 4). Again, the resolution does not allow us to determine how many sites were occupied at the same time. However, 376 sites with a total of 3828 dwellings in the area cover a timespan of c. 5000 years, suggesting that the number of sites and dwellings occupied at the same time was small and the population density low.

Figure 1: Fella is an example of a seascape with several habitation sites. Coastline at 10 m above present day sea level.

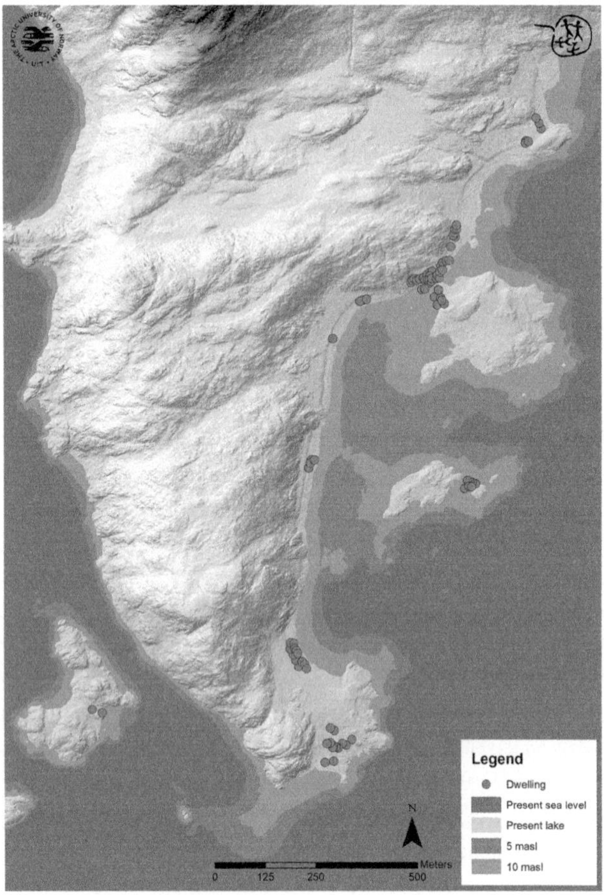

Map: M.S.Lindgren

The dynamics of residential units

The flexibility in households and residential units amongst foragers is often noted (e.g., Bird et al. 201; Bird-David 2017b). Direct evidence for the situation in prehistoric foraging societies is not possible to obtain. While it is highly problematic to employ ethnographic analogies, they do provide insights into recent practices. The following examples of demographic patterns are meant as illustrations of residential flexibility in sparsely populated coastal environments rather than direct analogies.

Along the coast of northwestern Greenland (with an outer coastline of roughly 600 km) 36 Inughuit winter sites were in use over the period 1910-1953. Ten to fifteen sites were in use each year, some almost every year, others only occasionally. The population numbered c. 250 persons and c. 60 households. The number of households at each site varied between 1 and 11, with an average of 4, and greater numbers were usually associated with trading posts. The number of inhabitants at each site typically varied between 10 and 30. Individual families rarely used the same winter site more than two years running, and the families co-residing changed constantly (Grønnow 2016).

Similarly, in the Ammassalik area in eastern Greenland, there are reports of 15 sites occupied in the winter of 1899-1900, each with only a single dwelling, but with an average of 27 inhabitants (and 403 in the district in total). An excavated dwelling measured 28 m^2. Many sites were occupied for only one year at a time. New and unrelated families could occupy the house for another winter, or spend the summer at the site, although families often returned to the same site or local area at regular intervals. Generally, related families chose to spend the winter together, but some families altered the relatives they resided with. Often relatives were also present at a neighbouring site (Møbjerg and Robert-Lamblin 1989)

The mid-Holocene Norwegian sites have several small contemporary dwellings, indicating the possibility of flexibility in the composition of the residential units. It is likely that a household relocated one or several times annually to access different seasonal resources. Such relocations may have been an opportunity for reconfiguring residential units, with the possibility of residing with different households seasonally, annually or at more irregular intervals.

Living in small residential units some distance apart over extended periods may have led to the type of very tight and intimate group dynamic described by Bird-David (2017b). Figure 2 is a visual representation from a specific archaeological site, which illustrates the closeness of the dwellings, and the intimate setting of daily life at the site. Regular alterations in compositions would have led to new constellations and enhanced the perception of being multi-relational. Burch provides an example from the Iñupiaq of northwest Alaska where four households make up a residential unit of 30 persons,

Figure 2: Artistic impression of life at the site Sundfjæra, based on the archaeological record.

©Endre Skandfer, Arctic University of Norway

but the total sum of all the individual family ties is an impressive 435 (Burch 2006: 101).

It is possible that under favourable weather conditions individuals or groups paid occasional visits to neighbouring residential groups in a nearby seascape. When visitors arrived, hosts did not necessarily perform other tasks, but they shared them with new persons (Bird-David 2017a: 214). This would have been a situation encouraging the exchange of knowledge and technological information. With regular restructuring of residential units and local visits, the close interaction and transmission of technological and socio-cultural knowledge would over the course of some years have extended to a much wider group than the c. 25 in a residential unit.

Regional networks

Indications of the size of the areas within which a residential unit obtained their resources, the frequency of residential relocation, and the distance of such residential changes/shifts is limited, while estimates for absolute population size in the region remain nonexistent.

Figure 3: A: T-shaped artefacts; B: animal headed daggers; C: Early Northern Comb Ware; D: Amber

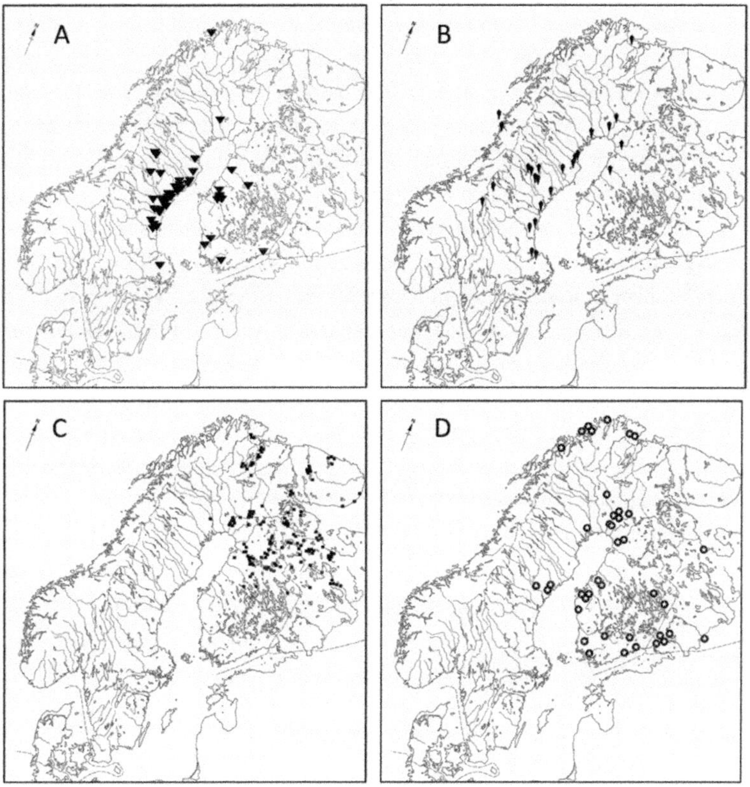

We do have indications for the existence of regional communities of practice. While there are overall similarities in the technologies and artefacts employed across northern Norway, northern Sweden, Finland, and Northwestern Russia (henceforth northern Fennoscandia), an area covering a total of c.1,000,000 km^2, there are also regional variations. Several distinct types of artefacts display a regional spatial dispersion. These include T-shaped slate tools concentrated along a c. 250 km coastal stretch of central northern Sweden (Figure 3A), slate daggers with animal heads found across 700 km along the coast of northern Sweden (Figure 3B), leaf-shaped bifacial chert points occuring over 700 km along the coast of northern Norway, and leaf-shaped slate points over at least 300 km along coastal northwestern Norway (Damm 2014). The first pottery, Early Comb Ware, was introduced from the east at c. 5300 BC and spread across eastern Fennoscandia but did not continue further west (Figure 3C). Analyses have indicated regional variation in decoration (Skandfer 2005). Each of these artefact types reveals a spatial range of interaction. The T-shaped tools show a clear fall-off pattern from the only available slate source in the centre of the distribution area, with several small concentrations up to 100 km away. This suggests a central location for manufacture, but also a spatial region within which such tools were in use. Across the Bay of Bothnia in Finland similar tools are found in smaller numbers but manufactured from local material. Copying the tool was in this case not difficult, but the small number of items in Finland suggests either that the use of the tool (i.e. the activity it was used for) was less frequent or that a different tool or tool material was used. To me this indicates different communities of practice. Similarly, the abrupt halt of the spread of pottery, where sites on one side of the Varanger fjord contain pottery and those on the other not, despite otherwise nearly identical inventories, show us that activities involving pottery were absent from one area, or performed differently there. It also indicates that different communities of practice (linked to objects such as pottery, slate tools and chert points) did not have the same spatial distribution. Instead, they only partially overlap, suggesting that they represent separate networks.

Regional network dynamics

In the fjord-sound system of western Finnmark, there are several clusters of seascapes (Figure 4). Similar clusters are found in neighbouring fjord-sound areas to the east and west. It is uncertain to what extent residential relocations incorporated adjacent fjords for resource exploitation. Given the natural geographical division in northern Norway, I would be inclined to suggest that the majority of the relocations happened within an area delimited by logistic maritime routes connecting seascape clusters, i.e. within western Finnmark. This area is c. 2,300 km^2 if restricted to land area, but 6,200 km^2 if marine resource areas such as fjords, sounds and the open sea are included. This corresponds to interpretations of contemporary data from western Norway, where technology, typology and raw material provenance studies suggest that primary interaction was concentrated in separate fjord-sound systems (Bergsvik 2006). Given the rather short distances in western Finnmark (80-100 km from the head of the Alta fjord to the seascapes on Sørøya) regular interaction between individuals, households and residential groups across the entire fjord-sound area is highly likely.

However, for both western and northern Norway there are also indications of interaction between such primary geographical areas, as demonstrated in the wider spatial distribution of provenanced lithic material, the technology of specific lithic points and distinct artefact types. This spread of technologies and material would have required individual or group visits to residential units in fjord-sound areas beyond western Finnmark, if not relocations for partners or other reasons.

Another possible basis for intra- or inter-regional interaction could be regular aggregation. Some northern groups historically had an annual pattern of aggregation and dispersal, often with larger groups at winter settlements followed by summer dispersal (e.g among Kets and the Sámi). We see no strong indications for such patterns in our area for the period in question. However, aggregations for shorter periods may have taken place regularly for collective foraging at seasonal peak resource concentrations, for exchange or for rituals and indeed often for a combination of purposes including social interaction and exchange of information. For mid-Holocene western Finnmark aggregation could have taken place in relation to the early summer fishing at primary salmon rivers (e.g. Alta river) or during the important early autumn reindeer hunting, where rock art scenes provide evidence for the use of corrals (Helskog 2012). Such aggregations could have been combined with ritual activities at the main area for rock art at the head of the Alta fjord.

There is a rich rock art record in Northern Fennoscandia from c. 5200 BC onwards. There are also significant similarities across this vast area in types of motifs (animals, humans, boats), in hunting and ritual scenes and in the incorporation of the micro-topography of the panel surfaces. However, the motifs are expressed with different stylistic templates and the dominant species vary between regions. Fennoscandia has more than 300 rock art sites, but only a small number of larger sites with many panels and motifs. These appear to be distanced 200-300 km apart (Gjerde 2018), possibly reflecting places for regional aggregation.

Aggregations would be occasions for kin and non-kin to meet and interact, for collaboration in hunting and fishing and later processing of the harvest, and for the transmission of related knowledge and technology in the process. Again, one must bear in mind that these transmissions took place within communities of practice and between persons participating in con-

Figure 4: Density of dwelling remains in Western Finnmark in northern Norway, based on 376 sites and a total of 3828 dwellings.

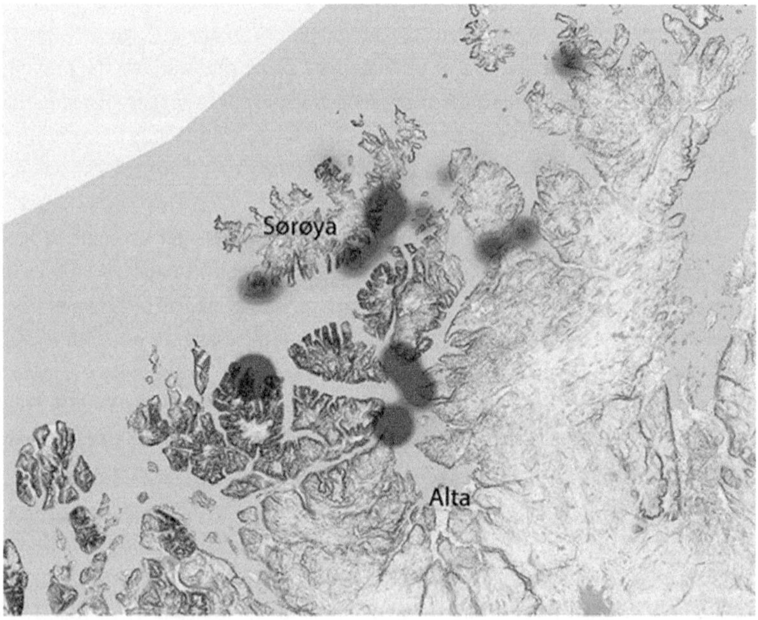

K.W.B.Vollan

crete activities. These communities may have been very open and inclusive but could for some tasks have been more narrowly delimited, thereby including a smaller number of skilled persons than the overall community and aggregated persons.

Long-distance interaction

Analyses of lithic adzes and axes deriving from known sources of volcanic greenstone at Lake Onega in Karelia show that the majority of preforms, indicating primary production, lie within 50 km from the source, and that beyond 150 km from the source there are only finished items. However, many such adzes and axes were found 200-700 km away (Tarasov and Nordqvist 2022). Similarly, the T-shaped slate artefacts (Figure 3A) cluster within a radius of c. 100 km from the known slate source, with some found up to 500 km away (Damm 2014). Other long-distance contacts are demonstrated through discoveries in northern Norway of amber beads from Latvia 1,500 km away (Figure 3D), a copper dagger from sources at Onega 1,000 km away, picks of material obtained at the head of Bothnian Bay 500 km away, and unique finds of animal headed daggers 500 km from their concentration in central northern Sweden (Figure 3B).

We have no concrete evidence (from for example isotope analyses) for journeys across 4-500 km. It is possible that for parts of these distances the objects were handed down the line from household to household. Nevertheless, recent research indicates that central parts of the inland (inner Finnmark) north of the Bothnian Bay were inhabited only to a limited extent in the mid-Holocene (Skandfer et al. 2022). In northern Sweden and Finland, habitation appears to be linked partly to the coast, but also partly to river and lake systems. The major routes of travel were therefore along the coast or linked to water systems, while crossing of watersheds appears to have taken place to a lesser extent. Hence, there were areas with limited activity and few occasions for passing on items. The items that did cross such natural geographies suggest that at least occasionally some persons or groups would journey longer distances to other regions bringing along goods and items (Damm & Skandfer 2022). That some persons for various reasons (adventure, exchange) travel longer distances is also known ethnographically, for example when large groups of umiaks travelled hundreds of kilometers northward along the west coast of Greenland, overwintering before returning to their home area the next summer – or small parties travelling from east to west Greenland often spending several years on the journey (Gulløv 1997, Jensen et al. 2011). Spending a season or more in a different region would provide not only exotic goods, but also new social relations, perhaps exchange partners, marriage partners, new information and stories and possibly new technological skills.

Intensity and extent of interaction

The key demographic entity in mid-Holocene northern Norway appears to be the small residential unit, consisting of a few households. Rather than view households and residential units as homogeneous faceless collectives, we must attempt to perceive the past as inhabited by a diversity of persons of different gender, age and capacities (e.g., Tringham 1991, French 2021). Also forager households are heterogenous. They consist of changing compositions of women and men, children, teenagers, adults and elderly, each with different competences, each involved in a different set of tasks. As a result, although they live in small residential units, a variety of patterns of interaction may be expected both within and beyond any current residential unit. This would have contributed to multiple situational identities or roles for each person, enhancing the multi-relational social fabric stressed by Bird-David (2017a, 2017b).

The most intensive interaction in mid-Holocene Norway would have taken place within the residential units, which we assume existed for many months. But if the co-residing households within these units were altered frequently, the number of persons one had close interaction with over time expanded significantly. With each new combination of households there was renewed potential for transmission of knowledge and skills within communities of practices.

The regional delimitations and variations in tools and rock art suggest that there were regional practice and learning networks, each producing and using distinct types. Such networks could only be established if there was regular interaction between individuals intra- and inter-regionally. Beyond the co-residing households, we may assume some informal and irregular contact between residential units, and possibly occasional or regular aggregation of many households for communal hunting/fishing and for social and ritual events. Such larger gatherings would have allowed for exchange of information and perhaps inspired more superficial copying, rather than transmission of underlying technologies and content of practices. It is possible that the activities performed varied between the small residential sites and larger aggregation sites. This difference in activities may have prevented transmission of details from practices not actively engaged in during larger gatherings. However, these events may have constituted the basis for new compositions of residential groups, or for a person to change household – through partnership, friendship or other types of motivations and obligations. Such flexibility was at the core of the social and cultural relations in such communities (see also Hofmann et al. 2016). Ultimately then, making new contacts at a gathering, which lead to relocation, could in turn lead to wider distribution of skills and knowledge. However, the quantity of people gathering at such an event may have had little bearing on the impact on knowledge transmission. While the number of people one person may have met over a lifetime may be substantial (as outlined by Bird et al. 2019), this does not necessarily reveal much about the closeness of the interactions. The number of people one individual had socially close relations to may have been significantly smaller.

It is not necessary for everyone within a regional community of practice to meet everyone else; the transmission may take place from one household to another as residential units alter or as new partners with other skills and experience join a household. While the present assumption is that the majority of mid-Holocene mobility took place at a regional geographical scale, spanning one or more fjord-sound geographical areas, there are also indications of long-distance contacts. This indicates a small-world network (Tarasov and Nordqvist 2022, Maier et al. this volume) where local and regional networking dominates, but where cases of long-distance interaction are evidenced.

On the other hand, extensive journeying was not always necessary for an individual to participate in a long-distance network. If one person relocated to another household for a season or longer, the entire receiving residential unit acquired a new social relation, with all his or her skills and knowledge. The incomers would perhaps not work closely with all members of the residential group. But if this was a prolonged stay, they would become integrated, and their knowledge gradually gain weight. In other words, one can stay put within a small geographical area, and meet a limited number of individuals, but acquire information as if one has travelled far.

Conclusion: Quantity and quality of encounters

Residential units of mid-Holocene maritime foragers in northern Norway were mostly small (<25 persons), and the majority of the population probably mainly exploited a limited coastal area covering one or two fjord-sound systems. However, if the composition of the residential units was flexible (as indicated by small dwellings of "household" size), a person could have lived with a much larger number of individuals than 25 in the course of their life. In addition, informal visits and regular aggregations would have considerably expanded the number of people any person would have met and interacted with directly. Furthermore, it is likely that some persons from the regional water system travelled farther away and returned with information and that long-distance travellers arrived for stays of some duration, if not permanently. In both cases, these perhaps rather few long-distance travellers would bring information from other regions to the persons staying put; even if they had long-distance interaction, although of a different kind. Considering interaction at several geographical scales, let us reflect upon the differences in impact regarding systems such as those for the transmission of technology and knowledge.

Flexibility in residential composition and mobility, and the resulting wider networks, allow news and technological innovation to spread. Whether or not residential units in northern Norway employed the information arriving with travellers must have depended on the extent to which travellers were integrated into a household or residential group. The spread of news and ideas partly depends on whom you trust and choose to imitate, suggesting, again, that the length and intensity of social interaction played an important role.

Scale matters! But while the assumed flexibility of residential units – in combination with regional mobility and aggregation – suggests that the number of interpersonal contacts over a lifetime was quantitatively high, their impact depended heavily on the quality and duration of those interactions. Beyond quantity, it is the intensity and quality of interaction between persons that truly matters.

Acknowledgements

The project "Stone Age Demographics: multi-scale exploration of population variations and dynamics" is funded by the Norwegian Research Council (grant no. 261760).

References

Apel, J. 2001. *Daggers, Knowledge and Power. The Social Aspects of Flint-Dagger Technology in Scandinavia 2350-1500 cal. BC.* Coast-to-Coast 3. Uppsala: Dpt. of Archaeology and Ancient History, Uppsala University.

Bergsvik, K.A. 2006. *Ethnic Boundaries in Neolithic Norway*. BAR International Series 1554. Oxford: Archaeopress.

Bird, D.W., R.B. Bird, B.F. Codding and D.W. Zeanah 2019. Variability in the organization and size of hunter-gatherer groups: Foragers do not live in small-scale societies. *Journal of Human Evolution* 131: 96-108.

Bird-David, N. 2017a. Before Nation. Scale-Blind Anthropology and Foragers´ Worlds of Relatives. *Current Anthropology* 58(2): 209-226.

Bird-David, N. 2017b. *Us, Relatives. Scaling and Plural Life in a Forager World.* Oakland: University of California Press.

Burch, A.S.jr. 2006. *Social Life in northwest Alaska. The Structure of Iñupiaq Eskimo Nations*. Fairbanks, University of Alaska Press.

Burke, A. 2021. Integrating Anthropological Theory and Computational Archaeology. In S. Gaudzinski-Windheuser and O. Jöris (eds.). *The Beef behind all Possible Pasts. The Tandem-Festschrift in Honour of Elaine Turner and Martin Street*. Pp. 287-295. Monographien des RGZM 157. Mainz.

Collignon, B. 2006. *Knowing Places. The Inuinnait, Landscapes and the Environment*. Circumpolar Research Series no. 10. Canadian Circumpolar Institute (CCI) Press.

Damm, C. 2012a. From Entities to Interaction: Replacing Pots and Peoples with Networks of Transmission. In *A Linguistic Map of Prehistoric Northern Europe*. R. Grunthal and P. Kallio (eds.). Pp. 41-62. Mémoires de la Societé Finno-Ugrienne 266. Helsinki: The Finno-Ugrian Society.

Damm, C. 2012b. Interaction within and between collectives: networking in Northern Fennoscandia. In *Networks, Interaction and Emerging Identities in Fennoscandia and Beyond*. C. Damm and J. Saarikivi (eds.). Pp. 125-138. Mémoires de la Societé Finno-Ugrienne 265. Helsinki: The Finno-Ugrian Society.

Damm, C. 2014. Interaction: When people meet. In *Northern Worlds – landscapes, interactions and dynamics*. H.C. Gulløv (ed.). Pp. 227-239. PNM Publications from the National Museum. Studies in Archaeology and history Vol 22. Copenhagen, University Press of Southern Denmark.

Damm, C. & M. Skandfer. 2022. A Neolithic Corridor between East and West. In Halinen, P., Heyd, V. & Mannermaa, K. (eds.). *An Ode to Mika*. Helsinki, Monographs of the Archaeological Society of Finland 10: 11-18.

Damm, C.B., M. Skandfer and P. Jordan 2022. Peopling Prehistoric Coastlines: Identifying Mid-Holocene Forager Settlement Strategies in Northern Norway. *Journal of Maritime Archaeology* 17: 131-160.

Damm, C.B., M. Skandfer, E.K. Jørgensen, P. Sjøgren, K.W.B. Vollan and P. Jordan 2020. Investigating Long-Term Human Ecodynamics in the European Arctic: Towards an Integrated Multi-Scalar Analysis of Early and Mid Holocene Cultural, Environmental and Palaeodemographic Sequences in Finnmark County, Northern Norway. *Quarternary International* 549: 52-64.

Engelstad, E. 1984. Diversity in arctic maritime adaptation. An example from the late Stone age of arctic Norway. *Acta Borealia* 2: 3-24.

French, J. 2021. *Palaeolithic Europe: A Demographic and Social Prehistory*. Cambridge: Cambridge University Press.

Gardner, P.M. 2019. Foragers with limited Shared Knowledge. In *Towards a Broader View of Hunter-Gatherer Sharing*. N. Lavi and D.E. Friesem (eds.). Pp. 185-194. Cambridge: McDonald Institute for Archaeological Research.

Gjerde, J.M. 2018. An overview of Stone Age rock art in northernmost Europe – what, where and when? In *Rock Art of the White Sea*. L.Sangmong and L.Janik (eds.). Pp. 204-225. South-Korea: Ulsan Petroglyph Museum.

Gosselain, O. 2000. Materializing Identities: an African Perspective. *Journal of Archaeological Method and Theory* 7(3): 187-217.

Gosselain, O. 2008. Mother Bella was not a Bella. Inherited and transformed traditions in southern Niger. In *Breaking Down boundaries. Anthropological Approaches to Cultural Transmission and Material Culture*. M. Stark, B. Bowser and L. Horne (eds.). Pp. 150-177. Tucson: University of Arizona Press.

Gosselain, O. 2010. Exploring the dynamics of African pottery cultures. In *The archaeology of Regional Technologies. Case studies from the Palaeolithic to the age of the Vikings*. R. Barndon, Ø. Engevik and I. Øye (eds.). Pp. 193-224. Lewiston N.Y.: Edwin Mellen Press.

Grønnow, B. 2016. Living at a High Arctic Polynya: Inughuit Settlement and Subsistence around the North Water during the Thule Station Period, 1910-53. *Arctic* 69, Suppl. 1: 1-15.

Gulløv, H. C. 1997. From Middle Ages to colonial Times. Archaeological and ethnohistorical studies of the Thule Culture in south west Greenland 1300-1800 AD. Monographs on Greenland/Meddelelser om Grønland: *Man & Society* 23.

Hallgren, F. 2008. *Identitet i praktikk. Lokal, regionala och överregionala sociala sammanhang innom nordlig trattbegerkultur*. Coast-to-coast 17. Uppsala: Dpt. of Archaeology and Ancient History, Uppsala University.

Helm, J. 1965. Bilaterality in the Socio-Territorial Organization of the Arctic Drainage Dene. *Ethnology* 4(4): 361-385

Helskog, K. 1984. The younger Stone Age Settlement in Varanger, North Norway. Settlement and population size. *Acta Borealia* 1: 39-70.

Helskog, K. 2012. Ancient depictions of reindeer enclosures and their environment. *Fennoscandia Archaeologica* XXIX, 29-54.

Hodgetts, L. 2010. Subsistence Diversity in the Younger Stone Age Landscape of Varangerfjord, Northern Norway. *Antiquity* 84(323): 41-54.

Hofmann, D., R. Ebersbach, T. Doppler and A. Whittle 2016. The Life and Times of the House: Multi-Scalar Perspectives on Settlement from the Neolithic of the Northern Alpine Foreland. *European Journal of Archaeology* 19: 596-630.

Jensen, E. L., K. Raahauge and H.C. Gulløv 2011. *Cultural Encounters at Cape Farewell. The east Greenlandic Immigrants and the German Moravian Mission in the 19th Century*. Monographs on Greenland vol. 348. Copenhagen: Museum Tusculanum Press.

Jordan, P. 2012. From hunter to Herder? Investigating the Spread of Transport Innovations in Northwest Siberia. In *Networks, Interaction and Emerging Identities in Fennoscandia and Beyond*. C. Damm and J. Saarikivi (eds.). Pp. 27-48. Mémoires de la Societé Finno-Ugrienne 265. Helsinki: The Finno-Ugrian Society.

Jordan, P. 2015. *Technology as Human and Social Tradition. Cultural Transmission among Hunter-Gatherers*. Oakland: University of California Press.

Jørgensen, E.K. 2020. The palaeodemographic and environmental dynamics of prehistoric Arctic Norway: An overview of human-climate covariation. *Quarternary International* 549: 36-51.

Kelly, R.L. 2013. The Lifeways of Hunter-Gatherers. The Foraging Spectrum. Cambridge, Cambridge University Press.

Lave, J. and E.Wenger. 1991. *Situated Learning: Legitimate Peripheral Participation*. Cambridge: Cambridge University Press.

Lundberg, Å. 1997. *Vinterbyar – ett bandsamhälles territorier i Norrlands inland 4500-2500 f.Kr.* Studia Archaeologica Universitatis Umensis 8. Umeå: Umeå Universitet.

Maier, A., I. Schmidt and A. Zimmermann. 2022. Putting numbers on "small-scale" – Observation on quantity, densities and connectedness of hunter-gatherers from an archaeological perspective. This volume.

Møbjerg, T. and J. Robert-Lamblin 1989. The settlement at Ikaasap Ittiva, East Greenland. An ethnoarchaeological investigation. *Acta Archaeologica* 60: 229-262.

Schanche, K. 1995. Var det store landsbyer i Finnnmark i yngre steinalder? Debattinnlegg med utgangspunkt i bosetningsspor fra ca. 2000 f.Kr. *Arkeologiske Skrifter 8, Steinalderkonferansen i Bergen 1993*. Pp. 174-186. Bergen: Universitetet i Bergen.

Sjögren, P.J. and C. Damm 2019. Holocene vegetation change in northernmost Fennoscandia and the impact on prehistoric foragers 12000-2000 cal. BP – A review. *Boreas* 48(1): 20-35.

Skandfer, M. 2005. Early, northern Comb Ware in Finnmark: The concept of Säräisniemi 1 reconsidered. *Fennoscandia Archaeologica* XXII: 3-23.

Skandfer, M., H.P. Blankholm and B. Hood 2022. *Archaeological Perspectives on Hunter-Gatherer Landscapes and Resource Management in Interior North Norway*. Sheffield: Equinox. In press.

Tarasov, A. and K. Nordqvist 2022. Made for exchange: the Russian Karelian lithic industry and hunter-fisher-gatherer exchange networks in prehistoric north-eastern Europe. *Antiquity* 96(356): 34-50.

Tringham, R.1991. Households with Faces: the Challenge of Gender in Prehistoric Architectural Remains. In *Engendering Archaeology. Women and Prehistory*. J. Gero and M. Conkey, M. (eds.). Pp. 347-373. Walnut Creek: Left Coast Press.

Vollan, K.W.B. forthcoming. Archaeological inferences under uncertainty: prehistoric dwellings, radiocarbon dates and the non-contemporaneity problem.

Wenger, E. 1998. *Communities of Practice. Learning, Meaning and Identity*. Cambridge, Cambridge University Press.

Whallon, R. 2006. Social networks and information: Non-"utilitarian" mobility among hunter-gatherers. *Journal of Anthropological Archaeology* 25: 259-270.

Comment by Elspeth Ready

The meaning and usefulness of the term "flexibility" as a descriptor of social organization has been debated in anthropology for a long time. In Inuit studies, flexibility refers to "the *prevalence* of situations in which no strong social preference is exerted, or even shown, for any one of several feasible courses of action" (Lange 1977: 107, emphasis added). For instance, no particular arrangement for Inuit post-marital residence was strongly prescribed and so it and might be described as a "choice," albeit one in which the conflicting desires and interests of many different people potentially played a role. In the Inuit case, flexibility in social organization emerges from the fact that situations in which active choice and consensus determine the course of action occur frequently and in multiple domains. In other contexts, flexibility in social organization may emerge from the possibility of choosing between several different culturally-specified alternatives (Aberle 1963), or from the fact that actual social arrangements do not match cultural models (Firth 1957).

Because different mechanisms can produce variable social organization, a problem with the term "flexibility" is that it is sometimes used to describe the variability in social organization itself and sometimes for the cultural traits that produce it. Wiessner (1982: 61) highlights this issue: "the apparent flexibility of organization among the !Kung [Ju/'hoansi] is not true flexibility in itself, but the product of a structured system of social organization." In this case "apparent flexibility" means observed variability, while "true flexibility" presumably means an absence of rules governing that organization.

Furthermore, simply using the term "flexible" does not explain variability in social organization (Cook 1966). Inuit studies have again had an important influence on functional explanations for flexible social organization (e.g., Willmott 1960): the lack of rules for deciding residential arrangements allowed group composition to be highly responsive to changing social and ecological opportunities and constraints at a fine temporal scale. As such, flexibility is often considered to be a cultural adaptation to scarce or unpredictable resources (Cook 1966). There are good theoretical reasons that this can be the case (e.g., Dyson-Hudson and Smith 1978), but observing variable social organization does not necessarily mean that that variation has an adaptive function.

I have cited old work here to highlight the time depth of debates about what flexibility is, but these issues are still relevant to understanding contemporary uses of the term and need to be considered when using the concept in archaeology. Although variation in social organization may leave material traces, the cultural practices that generate that variation are difficult to access through archaeological evidence alone. In contrast, the broader spatial and temporal scope of archaeology has advantages for examining potential adaptive explanations for variable group organization, for instance, in the ability to examine correlations between climate change and social organization over extended periods of time (e.g., Woollett 2007).

In her contribution to this volume, Damm infers "flexibility" in social organization from variability in the number of dwellings at residential sites and evidence for residential mobility (with different sites occupied seasonally and different locations potentially used from year-to-year). It is not clear to me that "flexibility" is more than a synonym for variability here, as variability across sites does not tell us much about the cultural mechanisms that produced it, nor if that variability was adaptive.

However, Damm's study tells us more interesting things about cultural processes at a different scale. Scaling out to regional patterns reveals the presence of different, partially overlapping networks ("communities of practice") for different tasks, as evidenced by, for example, lithic tool types or pottery styles. Thus, settlements were not little replicates of one another but rather need to be viewed as an interconnected system. From this coarse-grain view, the sets of practices that constitute "culture" did not come not as a single package bundle, but instead were locally assembled via persons who were part of communities of practice relating to different tasks. Despite their ultimate reliance on person-to-person interaction, these communities were likely not visible from the perspective of persons embedded within them.

This is interesting as each of these communities represented different kinds of knowledge, and may have been more or less tolerant of variability in practice or able to maintain fidelity of transmission through time. The structure of these networks, and the way in which different communities of practice overlapped, could undoubtedly could either enhance or constrain innovation or responses to change (Jones et al. 2021)—producing a different kind of "flexibility," or a lack thereof, at a much coarser scale than implied when ethnographers use the term (though I would not recommend proliferating uses of the word). I would be greatly interested to see future research exploring in more detail the spatial and temporal dynamics of these communities of practice.

References

Aberle, D. F. 1963. Some sources of flexibility in Navaho social organization. *Southwestern Journal of Anthropology* 19(1): 1-8.

Cook, E. A. 1966. Cultural flexibility: Myth and reality. *Anthropos* 61(3/6): 831-838.

Dyson-Hudson, R., & Smith, E. A. 1978. Human territoriality: An ecological reassessment. *American Anthropologist* 80(1): 21-41.

Firth, R. 1957. A note on descent groups in Polynesia. *Man* 57: 4-8.

Jones, J. H., Ready, E., & Pisor, A. C. 2021. Want climate-change adaptation? Evolutionary theory can help. *American Journal of Human Biology* 33(4): e23539.

Lange, P. 1977. Some principles of Inuit social interaction. In Paine, R. (ed), *The White Arctic: Anthropological Essays on Tutelage and Ethnicity*, pp. 107-128. St John's: Memorial University of Newfoundland.

Wiessner, P. 1982. Risk, reciprocity and social influences on !Kung San economics. In Leacock, E. and Lee, R. (eds.), *Politics and History in Band Societies*, pp. 61-84. Cambridge: Cambridge University Press.

Willmott, W.E. 1960. The flexibility of Eskimo social organization. *Anthropologica N.S.* 2: 48-59.

Woollett, J. 2007. Labrador Inuit subsistence in the context of environmental change: An initial landscape history perspective. *American Anthropologist* 109(1): 69-84.

Comment by Andreas Maier

When approaching archaeological questions on different spatial and temporal scales, the role of the individual is often a topic of concern. Charlotte Damm rightly points to the fact that all transmission processes, be it of skills or information, depend in the interaction between individuals. Individuals make up populations and are the basic acting unit that creates the archaeological record. Being thus undoubtedly a central player, the question arises whether individual actions and decision making is pivotal for all process scales, or whether there are instances when they become less relevant. Focusing on individuals makes sense at scales, where individual decision making can be meaningfully observed and has a major impact on the processes of interest, i.e., within the temporal and spatial action radius of the individual. For prehistoric societies, the temporal scale thus probably spans from moments up to several decades, rarely exceeding individual lifespans. Spatially, relevant impact will be largely restricted to the local and regional scale, rarely exceeding distance of 1000 km. Beyond these limits, the individual's potential for influencing processes is clearly reduced.

Many processes are scale-bound and not all processes are meaningfully observable at all spatial and temporal scales. This scale dependency applies for both quantitative aspects (size, extent, magnitude, frequency) and qualitative aspects (intensity, intimacy, trust, and content of encounters, network structure, network connectivity, social organization) of societies and social interaction. It also has a strong impact on system responses and feedback processes of social systems and governs the occurrence of emergent properties, i.e., characteristic and often decisive properties of a system which are observable only at certain scale levels, but not at others. With increasingly higher scale levels, the importance of individual decision making for the observable processes strongly decreases, while other factors are gaining in impact. Group behaviour, i.e., the emergent properties of group actions, not just the sum of the actions of individuals in that group, will become more important, alongside stochastic effects, for instance in the form of drift, accidental events, or long-term shifts in the environmental setting. It is therefore of major importance to match the process scale of interest with the scale of observation. If we are interested in individual decision making, the temporal and spatial scale of observation must be sufficiently small. If we are interested in large-scale processes, we must choose a large-scale perspective, where individuals and their actions are often no longer visible. Thus at certain scales, individuals and their conscious decisions are decisive. At others, individuals become invisible, and their decisions are just one factor among many others that are equally or even more important. Considering individual decision making and all its variation usually also means dealing with rather noisy signals. At higher scale levels, some signals may become less noisy. Therefore, observations on higher scale levels can provide additional information about social systems not observable at smaller scales and vice versa. Differences in observations at different scale levels are thus not necessary an expression of irreconcilable opposites, but rather can complement one another. This is illustrated by comparing the contribution of Damm to the one from Maier et al. Charlotte Damm argues from a temporally and spatially intermediate scale that is in congruent with the scale of individual lifetimes and sphere of action and informed by an exceptionally well-preserved archaeological site record. Maier et al., in contrast, argue from a temporally and spatially much larger scale, spanning several millennia and about 2 million square kilometres, thus surpassing individual action spheres by orders of magnitude. Nevertheless, both studies eventually agree that while being small in numbers and living predominantly in small groups, the investigated prehistoric foragers also had and maintained contact with many more people also over larger distances. Both findings are complementary inasmuch as they propose different network structures for their specific scale of observation. Charlotte Damm identified different and only partly overlapping network circuits as best fit for a regional spatial scale. Maier et al. argue for a structure that resembles a small-world network, where far-travelled individuals ensure the connectivity between different regional clusters. Together, these findings suggest that the network structure of foraging communities might be different at different scale levels. This has consequences for both the quality and quantity of transmission processes. In a structure, where many people are involved in passing information through the network, the amount of information that can be transmitted is higher than in a structure, where the transmission process depends on individuals. At the same time, the probability of copying errors (both beneficial and adverse) occurring 'on the way' is higher the more people are involved.

Eventually, there can be no general claim to the necessity to engage with the sphere of individuals in archaeological research, nor can it be dismissed as unimportant. The question whether prehistoric research should be concerned with individuals or if "faceless collectives" are just as fine or even better is more than a personal preference: It is a matter of scale.

A large-scale view on 'small-scale societies'

Andreas Maier, Isabell Schmidt, and Andreas Zimmermann

Introduction – Quantitative and qualitative aspects of 'small-scale societies'

In ethnographic and archaeological research, the term 'small-scale society' has gained some popularity, in particular with regard to Palaeolithic communities. The usage of this label ranges from a synonym for Palaeolithic foraging societies across cases designating smallness in numbers (e.g., Jordan et al. 2013) or being restricted to a small area, to living in a local, kinship-based interaction network (e.g., Firth 1951) or having a non-centralized political system (e.g., Spielmann 2002; see Reyes-García et al. 2017 for a short overview). There are also combinations of several of these meanings. But recently, the notion of hunter-gatherer societies as being small-scale in the qualitative sense of living in kinship-based interaction networks of nested communities has been challenged, considering that while being perhaps small in population size, people are nonetheless living in fluid and large-scale social networks (Bird et al. 2019).

Generally speaking, scale levels (e.g., small – medium – large/local – regional – global) are used to refer to both quantitative and qualitative properties of objects, processes, or systems. In their quantitative sense, they convey relative notions about the size, extent, magnitude, or frequency of the investigated phenomena. In their qualitative sense, however, they also convey statements about properties of systems that are bound to and therefore characteristic for specific scale levels. This scale dependency of properties governs specific feedback processes, timing of system responses to external factors, and the occurrence of so-called emergent properties, i.e., characteristics of a system only observable at certain scale levels, but not at others (Zhang et al. 2004). It follows that not all questions can be addressed meaningfully at all scales and that it is necessary to match the process scale(s) of interest with the scale(s) of observation. In this contribution, we explicitly report observations from a large-scale perspective with time frames covering several millennia each, and a spatial extent of roughly two million square kilometres. We present estimates on the number, density, and connectedness of Upper Palaeolithic hunter-gatherers in Europe between 43,000 and 15,000 years (ka), thus addressing three fundamental aspects of 'small-scale' societies. We also explore how diachronic changes in these three factors – number, density, and connectedness – affect the evolution of material culture.

Instead of focussing on processes that operate over the lifespan of people, we target a much higher temporal scale level, where individuals turn into sometimes criticized 'faceless' collectives (French 2021; Damm, this volume). As a result, our findings can and probably will differ from observations at smaller scales. However, in light of what is stated above, we think that faceless collectives can contribute meaningfully to the question to what extent and in what respect hunter-gatherers are living in small-scale societies. We therefore see possible differences between our conclusions and those drawn from analyses at different scales as complementary rather than as conflicting, because: scale matters.

A short history of Upper Palaeolithic population and network development in Western and Central Europe

The quantitative aspects presented below are the results of palaeodemographic estimates carried out following the Cologne Protocol, an algorithm which provides regionally differentiated numbers and densities for mobile and sedentary societies (Schmidt et al. 2021). Inferences concerning qualitative aspects, namely the connectivity of the interaction networks, are based on similarities and differences in material culture traits. Fundamental here is the assumption that interaction between individuals and groups fosters significant similarities in the archaeological record (Boyd and Richerson 1985), while in the case of isolation already the phenomenon of drift will likely cause the accumulation of regional idiosyncrasies (Neiman 1995). Since 'drift is a consequence of sampling, it is amplified in smaller populations in which the number of people to copy from, and the number of objects or traits to copy are limited' (Buchanan and Hamilton 2009, 280). It follows that with a low network connectivity, difference in the material record will likely increase and overall similarity will decrease (Shennan 2000, 2001; Henrich 2004). We fully agree with Damm (2012a; this volume) that attention should be given to the fact that some similarities arrive easier than others and that different manufacturing processes can result into morphologically similar results. This is particularly true for studies concerned with processes that operate on small and intermediate spatial and temporal scales and are interested in individual decision making. On a large scale, however, the multitude of signals from the noisy choir of individual decision making are no longer observable. Averaged over millennia, individual actions cancel each other out or amplify one another, but eventually tune into a large-scale trajectory. This sum of individual decision making is not only a result of conscious actions, but also of accidental events, transmission errors and stochastic effects (Rindos 1989). Therefore, from the temporally large-scale perspective taken in this contribution, individual decision making is but one of many factors contributing to the observable processes and is thus not of major relevance.

Network formation: 43,000 to 33,000 years ago

At around 43,000 years ago, Anatomically Modern Humans had spread over large parts of Europe (Cortés-Sánchez et al. 2019). However, the population was not evenly distributed across the continent. On the contrary, in this period people were living in several regional clusters, in the following referred to as 'Core Areas', spatially separated from one another by areas which were only ephemerally used or totally uninhabited (Schmidt and Zimmermann 2019). Taken together, the Core Areas covered about 104,000 km² with a population estimate of presumably around 1,500 people living at the same time (Table 1). The average population density within these Core Areas has been estimated to about 1.5 people per 100 km² (Schmidt and Zimmermann 2019). The lithic and osseous tools during this period are remarkably similar throughout the area of investigation, while personal ornaments from shells (Vanhaeren and d'Errico 2006) and procurement areas of lithic raw materials (Schmidt and Zimmermann 2019) show regional differences. These findings indicate that regional communities in Franco-Cantabria, the Rhine-Meuse Area, the Upper Danube, and around the Western Carpathians (archaeologically visible via Core Areas, personal ornaments, and raw material procurement) maintained a highly efficient communication network among one another, spanning at least 2000 km from east to west and 1000 km from north to south (archaeologically visible via the strong similarities in lithic and osseous technology and tool design). However, a surprisingly clear boundary with regard to personal ornaments can be found that roughly coincided with the eastern border of present-day Germany, where – despite large overlaps with other groups – sites in Germany and Austria have a mutually exclusive spectrum of adornments (Vanhaeren and d'Errico 2006).

Network densification: 33,000 to 29,000 years

Within the next 5,000 years, roughly until 29,000 years ago, we can observe the growth and the emergence of new Core Areas alongside population growth and a densification of the network. The total extent of the Core Areas more than doubled to roughly 243,000 km² and the average amount of people living at the same time almost doubled to 2,800 people (Maier and Zimmermann 2017). At the same time, the density of people within the Core Areas dropped slightly to about 1.2 people per 100 km², an observation in accordance with the expansion of the population into previously uninhabited areas. However, this expansion process did not coincide with growing distances between the Core Areas or a thinning of the large-scale spatial structure of the network. To the contrary, during this period, a new Core Area forms in a geographic key region, namely around the Burgundy Gate (Maier et al. accepted, Fig. 9b). This is the only region in Europe, where three large rivers spring from relatively nearby sources, but flow in three different directions: The Rhone to the south, the Rhine to the north, and the Danube to the east (Maier 2019). Assuming that larger rivers served as important landmarks for long-distance travel (Hussain and Floss 2016), this area has high potential to form an important hub in the large-scale communication network at that time. An effective flow of information throughout the network from the Atlantic coast to the East European Plain – and thus a high connectivity – becomes evident in striking morphological similarities in female figures (Gaudzinski-Windheuser and Jöris 2015). Two specimens from Willendorf, Austria, and Kostenki, Russia, for instance, show almost identical traits despite a distance of about 1,700 km. Besides these overarching similarities, medium-scale differences are also observable. Regarding the lithic and osseous projectiles, for instance, a division in a western and eastern part of the network becomes apparent. The boundary between both parts still roughly coincides with the eastern border of present-day Germany, already observable during the previous period. The western part includes the area up to the Atlantic coast, while the eastern part extends eastward. These differences are reflected in the names of the archaeological units in both areas. The western assemblages are subsumed under the term Gravettian, while those in the eastern part are referred to as Pavlovian.

Table 1: Palaeodemographic estimates for the Upper Palaeolithic of Europe

Period in ka	**Core Areas (CA) in km2**	**Population size**			**Averaged density in CAs**
		min	*median*	*max*	*per 100 km²*
42-33	103,686	880	1,550	3,800	1.5
33-29	243,039	1,660	2,760	3,610	1.2
29-25	123,810	660	1,000	1,530	0.8
25-20	275,413	1,330	3,240	6,260	1.2
20-15	332,949	4,820	7,600	10,520	2.6

Network disintegration and fragmentation: 29,000 to 25,000 years ago

In sharp contrast to the previous period, the time between 29,000 and 25,000 years ago is characterised by a population decline both in numbers and distribution. The extent of the Core Areas shrank to 124.000 km² – but only half the area of the previous period – and the average population density within the Core Areas decreased to 0.8 people per 100 km². The estimated average number of people living at the same time dropped to 1000, only about one third of the previous period and probably close to the threshold of a minimal viable population (Maier and Zimmermann 2017). This decline affected the northern mid-latitudes particularly strong, leading to the disappearance of Core Areas north of the Alps. The Core Area in the Burgundy region, which had emerged in the previous phase, shrank strongly and the Core Area in the Upper Danube Area disappeared entirely. However, a decline in the number of people is observable for virtually all regions in Europe, indicating regional population breakdowns rather than movements of people from the north to the south (Ibid.). The decline in population and abandonment of large parts of Central Europe coincided with the disintegration of the large-scale network and fragmentation into two smaller structures. In consequence, regional idiosyncrasies accumulated within both networks. North of the Alps, the rupture in the large-scale, long-distance network followed again roughly the border observed for previous period. The western network contracted markedly to areas west of the Rhine, while the eastern network, referred to as Willendorf-Kostenkian, roughly kept the overall spatial extent.

Network reorganisation: 25,000 to 20,000 years ago

The trend in population decline and range contraction stopped between 25,000 and 20,000 years ago. This period is connected to a drastic expansion of the total extent of the Core Areas to 275,000 km – surpassing even the extent of the previous expansion between 33,000 and 29,000 years – and a strong increase of the population density within the Core Areas to 1.2 people per 100 km^2. The average amount of people living at the same time more than tripled in comparison to the previous period to 3,200 people per 100 km^2. This population increase, however, is only observable for Western Europe, while the population in Central Europe remains at very low levels (Maier et al. 2016). This strong imbalance already speaks in favour of two separated networks with no or very low contact between one another. This impression is corroborated by the accumulation of regional idiosyncrasies that started in the previous period and now become particularly pronounced. Between 25,000 and 20,000 years, differences in material culture between areas east and west of the Alps are probably the strongest throughout the entire Upper Palaeolithic in Europe. Roughly between 25,000 and 23,000 years, sites in southern France and on the Iberian Peninsula show a very characteristic and regionally differentiated artefact spectrum referred to as Solutrean (Schmidt 2015), not shared by other sites outside this area. East of the Alps, assemblages contemporaneous to the Solutrean also seem to reflect a shared technological and typological background with regional differences. It is interesting in this regard that the Core Area in Burgundy still does not appear again. However, this state of two largely separated networks was overcome again already between roughly 23,000 to 20,000 years ago. At around that time, assemblages occur in Western Europe which are referred to as Badegoulian and which bear close resemblance to contemporaneous assemblages in Central Europe (Ducasse et al. 2021; Händel et al. 2021).

Network reinvigoration: 20,000 to 15,000 years ago

Between 20,000 and 15,000 years ago, the population grew again in both areas, Western and Central Europe. Previously abandoned areas became repopulated, and people expanded further north. During this period, the estimated amount of people living at the same time reaches maximum values for the Upper Palaeolithic with a median estimate of 7,600 people (Kretschmer 2015). The total extent of all Core Areas rose to 333,000 km^2 and population density within the Core Areas more than doubled to 2.6 people per 100 km^2. It is dur-

Figure 1: A temporally large-scale view on diachronic change in population size and extent of Core Areas (Schmidt et al. 2021), network size and connectivity (estimated according to the spatial distribution of Core Areas and similarities/dissimilarities in the archaeological record) and artefact diversity (estimated from a coarse diachronic survey and the data from Maier et al. 2021b) plotted against solar summer insolation at 60°N (as implemented in CalPal-Beyond the Ghost, Version 2016.2, Weninger et al., 2014) as a proxy for the timing and productivity of the vegetation period (cf. Maier et al. accepted).

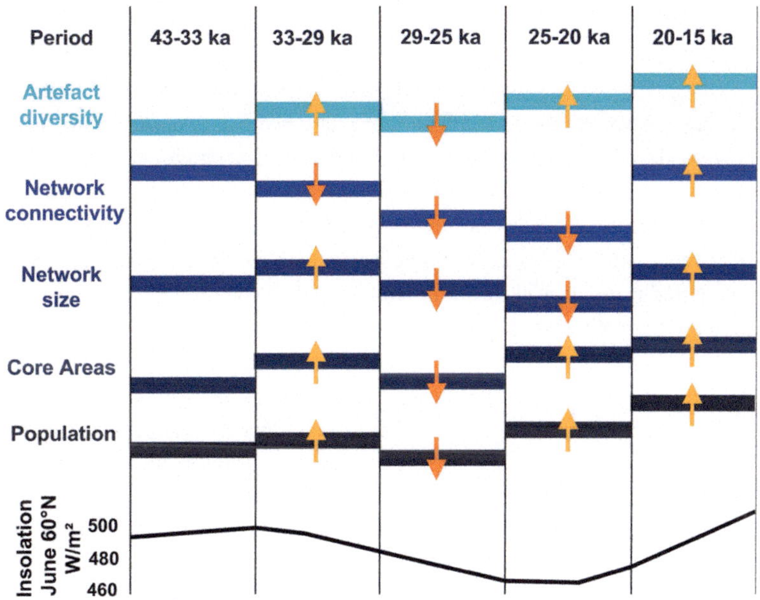

ing this time that the areas of the Burgundy Gate and the Upper Danube Valley become re-integrated in the settlement pattern, probably an important cornerstone in the re-establishment of long-distance communication patterns (Maier et al. 2020). With regard to the material culture record, overarching similarities found from the Atlantic coast up to the Dnieper River indicate that ideas were circulating again on a pan-European scale (Gaudzinski-Windheuser and Jöris 2015). The re-established communication network must have been very efficient. Its presumably high connectivity can be observed at the advent of a special kind of hunting equipment, so-called barbed points, which occur virtually simultaneously in the Pyrenees and the Carpathians at around 16,000 years ago (Maier et al. 2020). However, within the large-scale network, the previous interaction structures are still visible. On a medium scale, two sub-ordinate networks are observable. The western network is visible through a far-flung pattern of mollusc transport, spanning from the Atlantic and Mediterranean over the Paris Basin to the Rhine valley. The eastern network, in contrast, does not participate in this pattern (Maier 2015). The border between these two medium-scale networks roughly runs from the Herzynian Mountains over the Bavarian Forest to the Alps, thereby following a course surprisingly similar to the border observable in previous periods.

Discussion

The brief survey of population dynamics and network development during the Upper Palaeolithic highlights that small in numbers, restricted to a small area, and living in local networks are three aspects of being 'small-scale' that are largely independent from one another and do not necessarily co-vary. Indeed, relatively many people can live in networks with comparatively small spatial extent and comparatively pronounced regional idiosyncrasies, as seems to be the case between 25,000 and 20,000 years ago in Western Europe. By contrast, relatively few people can maintain comparatively large networks with a high connectivity (Bird et al. 2019), as seems to be the case between 43,000 and 33,000 years ago (Fig. 1). With regard to networks, it is between 29,000 and 25,000 years ago that Palaeolithic communities were probably at their 'smallest scale' since the arrival of Anatomically Modern Humans in Europe. Being small in numbers, densities and distribution and living in a social environment of network disintegration, all of the discussed parameters were in a 'small-scale' state (Fig. 1). However, even then people were not living in truly 'small-scale' networks, since contact can still be traced over larger distances, connecting several Core Areas.

What reduces the network scale of Upper Palaeolithic societies?

Looking at the archaeological record, it seems that Upper Palaeolithic hunter-gatherers sought to build and maintain large-scale networks, if possible. It is important to stress that the archaeological evidence for these far-flung networks is not the result of the construction of palimpsest by averaging the behavioural patterns of many small-scale communities from different periods over large time frames. Evidence that these networks were active during the lifetimes of individuals are shown by objects transported over 800 km from the Atlantic coast to the Rhine (Maier 2015) or the quasi-simultaneous adaption of technological novelties in Western and Central Europe that happened below the resolution of modern AMS radiocarbon dating, i.e., some decades (Maier 2020). The question thus arises: What internal and external factors can trigger downsizing processes in the different aspects of hunter-gatherer societies and how do they influence one another?

Here it is important to point out that this question can be asked on different temporal scales. Depending on the scale of observations, different factors must be considered. For instance, decision making of individuals and groups (i.e., choosing one option over others by reflecting available arguments) and traditions (i.e., choosing one option over others by usually unquestioned routines) surely have the power to influence the scale of networks (Codding et al., this volume). Decisions or traditions against interaction with others will have downsizing effects on the network scale. They can thus leave traces at higher scale levels, but have their main effects in shorter periods of several decades or centuries. The temporal scale on which these factors have their main impact is thus usually much smaller than our five time frames of several thousand years each. At such a large temporal scale, decision making becomes but one factor. Other factors, in turn, become more important. In the following, we therefore focus on factors which are better observable at large temporal scales, namely environmental change, size and distribution of populations, size and connectivity of networks, and artefact diversity.

Given that Upper Palaeolithic hunter-gatherers had little influence on climate and seasonality, we start our reflections with these parameters external to the human system. It has been found that there is an interesting correlation between changes in solar insolation and long-term trends in population dynamics (Maier et al. accepted; Fig. 1). The reason for this might be that solar insolation has an influence on the timing and productivity of the vegetation period, which in turn influences migrating animals and thus resource availability for Upper Palaeolithic hunter-gatherers. Long-term trends in resource availability eventually seem to have visible impact on the size, density and distribution of hunter-gatherer populations. The observable long-term demographic trends, in turn, seem to have a strong influence on the long-term development of the extent and connectivity of networks. Looking at the Upper Palaeolithic record, it seems that population decline coincides with network disintegration and fragmentation. It is, however, noteworthy that during the period between 33,000 and 29,000 years ago network connectivity apparently declined, while the size and distribution of the population as well as the extent of the network grew.

In this regard it is interesting to consider that the connectivity of a network can be negatively influenced not only by too few participating individuals, but also by too many. Although it has been demonstrated that there is no fixed upper limit for human contacts defined by the size of their neo-cortex (Lindenfors et al. 2021), as has been suggested by Dunbar (1992), keeping contact requires the investment of time and energy. The same goes for learning and teaching skills. As a consequence, above a certain number of groups in the network, no group can maintain constant exchange with all other groups, and indirect contacts of 2nd, 3rd, etc. order will increase. Moreover, with more people inside each group, the necessity for maintaining long-distance networks for mating or subsistence security decreases. The maximum number of direct contacts between groups might be very specific for certain periods and areas, since it depends on the number of people per group, the geographic distribution of populations and topographic barriers, social mobility rules, the number of cultural traits available for learning, as well as transport and communication technology. In this regard, larger populations might even show a tendency to form spatially less extensive networks than smaller ones.

While low population numbers can thus be insufficient to maintain large-scale networks, high numbers can decrease the necessity to do so or exceed the available energy for networking activities with all members. In both cases, network connectivity will decrease, and regional idiosyncrasies (the inverse function of network connectivity) will increase. However, the increase in regional idiosyncrasies (decrease in connectivity in Fig. 1) during the Upper Palaeolithic is of two different kinds, one is related to the increase of artefact diversity, the other one to its decrease. Artefact diversity during the Upper Palaeolithic, in turn, seems to be strongly positively correlated with population size (Fig. 1). Looking at the archaeological record, it thus seems that the decrease in network connectivity between 33,000 to 29,000 years and 25,000 to 20,000 years ago is related to an increase in artefact diversity linked to increasing population size and distribution, maybe exceeding the networks specific capacities. Between 29,000 and 25,000 years ago, by contrast, decreasing connectivity is related to a loss in artefact diversity, linked to strong population decline and network disintegration. While the rising regional idiosyncrasies between 33,000 and 29,000 years ago are thus 'differences of affluence', those between 29,000 and 25,000 years ago are 'differences of loss'.

Small-scale vs. small-world

Reflections on connectivity also raise questions about the structure of Upper Palaeolithic hunter-gatherer networks and how a high connectivity can be maintained at a pan-European scale, when the total number of individuals is small, and most individuals spend most of their lives interacting with others from the same region. The number of personal contacts and the area covered during a lifetime can differ markedly (Damm, this volume; Codding et al., this volume). However, it can be stated that the interactions of an individual throughout a lifetime is finite and unlikely to cover all other individuals living in the same network at a supra-regional or continental scale.

On an intermediate, regional spatial scale, networks can consist of several task-specific sub-network, or 'circuits', within which similar but not identical groups of people interact. These circuits are not nested, but broadly overlap with one another (Damm 2012b). Such a network structure would be rather robust against external distortion, since its connectivity does not rely on individuals, but is ensured by many members in broadly redundant circuits.

However, while such a network structure works on a regional scale, where distances between most members can be travelled within a few days, it is rather unlikely for large-scale and far-distance contacts, since the energy investment beyond certain distances would drastically exceed the benefits. Here, so-called small-word networks (Milgram 1967) offer an interesting model. Networks can be described in terms of the nature of contact between individuals from local (direct contact only with neighbouring individuals) to random (direct contact with potentially everybody in the network) and the corresponding path-length, i.e., the number of individuals that is needed to pass an information from one side of the network to the other (Watts and Strogatz 1998; Bentley and Maschner 2008). In contrast to local networks, which are characterized by exclusively local connections and therefore a high path-length, random networks are characterized by many cross-cutting connections between individuals and thus a shorter path length. In small-world networks, however, most individuals have only local contacts, but few individuals have long-distance contacts. These few long-distance contacts reduce the characteristic path-length of the network tremendously, almost to the extent of random networks. For Palaeolithic societies, this means that few far-travelling individuals can provide long-distance contacts between otherwise mainly regionally cantered groups, thereby lowering the characteristic pathlength of the network significantly, enhancing its connectivity and ensuring an effective flow of information throughout the entire network (Bentley and Maschner 2008.). Relying on few individuals to connect many, such networks can show a low resilience to distortion. A decline in population with decreasing numbers of far-travelling individuals may severely impact the connectivity of such networks. Erasing important hubs from the network, as can be those in Burgundy or the Upper Danube area, easily leads to a fragmentation of the network into several units of smaller scale.

Concluding remarks

There are three main conclusions following from our reflections. First, Upper Palaeolithic hunter-gatherers in Western and Central Europe were living in closely integrated communities of small size. This does not imply that they were or were not *per se* small-scale communities. Rather, it seems that some aspects of small-scale societies can be observed, for instance that they were very low in numbers. Other observable aspects, however, are at odds with the notion of small-scale societies. Evidence for long-distance interactions during the lifetime of individuals show that Upper Palaeolithic foragers actively maintained large-scale networks. Maintaining networks with a high long-distance connectivity is a good strategy to mitigate negative effects intrinsic to small groups, such as random variation in demographic parameters, inbreeding, as well as the loss of cultural knowledge because of drift, for instance.

Second, there seem to be differences between network structures, depending on the process scale. On a regional scale, overlapping circuits of changing composition seem to be a plausible assumption. Because of many redundant connections, regional networks are comparatively stable and resilient to distortions. On a large scale, in contrast, the network structure might have been rather like small-world networks, where few individuals travelling between close-knit regional groups ensured long-distance communication. Such a network structure would have much less redundant connections and thus would be more likely to disintegrate in case of distortions.

Third, size and connectivity of the large-scale, long-distance networks seem to be strongly dependant on demographic thresholds. From a temporally large-scale perspective, demographic long-term developments, in turn, seem to be coupled to environmental change. Depending on social mobility rules and the available transport and communication technology, these networks seem to have had historically contingent conditions when the flow of information through the network connectivity was optimal, i.e., all members had access to all information virtually at the same time. Such optimal conditions would have a strongly homogenizing effect on the archaeological record and are potentially part of an explanation for the strong similarities between 43,000 and 33,000 years ago. There are two ways how large-scale network connectivity can deviate from such optimal conditions. The first is population decline alongside habitat fragmentation and eventually large-scale network disintegration. This process, observable between 29,000 and 25,000 years ago, seems also to be connected to decreasing artefact diversity. Given that this decrease is likely to exhibit stochastic properties because of drift phenomena differing within disconnected regions, the process thus fosters the formation of regional 'differences of loss'. At the same time, network fragmentation due to population decline is connected to the danger of becoming truly small-scale and thus perceptible to the perils connected with it. The second way of deviating from optimal connectivity starts from increasing population size and distribution. On the one hand, having more people in the regional neighbourhood reduces the necessity to maintain long-distance networks for mating, subsistence, etc. On the other hand, having many people in your local and regional network might exhaust the temporal capacity of individuals to maintain personal contacts beyond the regional scale. As a consequence, the network's large-scale connectivity and thus the homogenous flow of information throughout the entire network will decline, while regional sub-networks, or 'circuits' become stronger within which information on certain cultural traits circulates. This process thus fosters increasing artefact diversity, although the occurrence of certain traits is restricted to specific areas of the network, causing 'differences of affluence'. Such processes might be observable in the archaeological record for instance between 33,000 and 29,000 years ago. Accumulating regional idiosyncrasies which do not arise from stochastic loss, as in the former, but from innovations, as in the latter case, are beneficial for the innovative potential of a community. An optimal flow of information throughout the entire network that homogenises differences in artefact diversity might thus not be optimal for the technological development, since it can counteract the accumulation of cultural traits in different regions, thereby lowering the material for potential innovations. Such a reading of the archaeological record is also in line with the finding from a computer-based experiment on the accumulation of innovations (Derex and Boyd 2016). Disruptions in the connectivity of large-scale networks due to population decline thus seems to put societies in danger of becoming truly small-scale with negative effects for their viability and technological development, while a reduced connectivity due to population increase can be beneficial, since it fosters the accumulation of regional artefact diversity. However, a reduced maintenance of long-distance contacts again raises the danger of disruption in periods of crisis.

Acknowledgements

We are grateful to the organisers for inviting us to the interesting workshop 'Scale Matters: The Quality of Quantity in Research on Social Relations'. Thanks go also to our reviewers for their constructive criticism. The Palaeodemographic research presented was funded by the Deutsche Forschungsgemeinschaft, CRC 806, 'Our Way to Europe', Project-ID 57444011.

References

Bentley, R. A., and Maschner, H. D. G. 2008. Complexity theory. In R. A. Bentley, H. D. G. Maschner, and C. Chippindale (eds.), *Handbook of archaeological theories* (pp. 245-270). Lanham, MD: Rowman and Littlefield.

Bird, D. W., Bliege Bird, R., Codding, B. F., and Zeanah, D. W. 2019. Variability in the organization and size of hunter-gatherer groups: Foragers do not live in small-scale societies. *Journal of Human Evolution* 131, 96-108.

Boyd, R., and Richerson, P. J. 1985. *Culture and the evolutionary process*. Chicago: University of Chicago Press.

Buchanan, B., and Hamilton, M. J. 2009. A formal test of the origin of variation in North American early Palaeoindian projectile points. *American Antiquity*, 74 (2), 279-298.

Codding, B. F., Cole, K., and Wilson K., M. (this volume). Socioecological Factors Influence Hunter-Gatherer Group Size, Lifetime Interactions, and Emergent Properties of Culture.

Cortés-Sánchez, M., F. J. Jiménez-Espejo, M. D. Simón-Vallejo, C. Stringer, M. C. Lozano Francisco, A. García-Alix, J. L. Vera Peláez, C. P. Odriozola, J. A., Riquelme-Cantal, R. Parrilla Giráldez, A. Maestro González, N. Ohkouchi, and Morales-Muñiz, A. 2019. An early Aurignacian arrival in southwestern, Europe. *Nature Ecology & Evolution* 3 (2), 207-212.

Damm, C. (this volume). Scales of interaction. Quantity and quality of encounters amongst northern foragers.

Damm, C. 2012a. From Entities to Interaction: Replacing Pots and Peoples with Networks of Transmission. In *A Linguistic Map of Prehistoric Northern*

Europe. R. Grunthal and P. Kallio (eds.). Pp. 41-62. Mémoires de la Societé Finno-Ugrienne 266. Helsinki: The Finno-Ugrian Society.

Damm, C. 2012b. Interaction within and between collectives: networking in Northern Fennoscandia. In *Networks, Interaction and Emerging Identities in Fennoscandia and Beyond*. C. Damm and J. Saarikivi (eds.). Pp. 125-138. Mémoires de la Societé Finno-Ugrienne 265. Helsinki: The Finno-Ugrian Society.

Derex, M., and Boyd, R. 2016. Partial connectivity increases cultural accumulation within groups. *Proceedings of the National Academy of Sciences* 113, 2982-2987.

Ducasse, S., Chauvière, F.-X., Pétillon, and J.-M. 2021. Breaking bad? Discarding the solutrean norms: Chronology, evolution and geographical extent of the badegoulian phenomenon in Western Europe. *Quaternary International*, 581-582, 61-83.

Dunbar R. I. M. 1992. Neocortex size as a constraint on group size in primates. *J. Hum. Evol.* 22, 469-493.

Firth, R. W. 1951. *Elements of Social Organization*. Watts, London, UK.

French, J. 2021. *Palaeolithic Europe: A Demographic and Social Prehistory*. Cambridge: Cambridge University Press.

Gaudzinski-Windheuser, S., and Jöris, O. 2015. Contextualising the female image – symbols for common ideas and communalidentity in Upper Palaeolithic Societies. In F. Coward, R. Hosfield, M. Pope, and F. Wenban-Smith (eds.), *Settlement, Society and Cognition in Human Evolution: Landscapes in Mind* (pp. 288-314). Cambridge: Cambridge University Press.

Händel, M., Simon, U., Maier, A., Brandl, M., Groza-Săcaciu, S. M., Timar-Gabor, A., and Einwögerer, T. 2021. Kammern-Grubgraben revisited – First results from renewed investigations at a well-known LGM site in East Austria. *Quaternary International* 587, 137-157.

Henrich, J. 2004. Demography and cultural evolution: How adaptive cultural processes can produce maladaptive losses – the Tasmanian case. *American Antiquity*, 69 (2), 197-214.

Hussain, S.T. and Floss, H. 2016. Streams as Entanglement of Nature and Culture: European Upper Paleolithic River Systems and Their Role as Features of Spatial Organization. *Journal of Archaeological Method Theory* 23, 1162-1218.

Jordan, F. M., C. van Schaik, P. Francois, H. Gintis, D. B. M. Haun, D. H. Hruschka, M. A. Janssen, J. A. Kitts, L. Lehmann, S. Mathew, P. J. Richerson, P. Turchin, and P. Wiessner. 2013. Cultural evolution of the structure of human groups. Pages 88-116 in M. H. Christiansen and P. J. Richerson, (eds.). *Cultural Evolution: Society, Technology, Language, and Religion*. MIT Press, Cambridge, MA.

Kretschmer, I. 2015. *Demographische Untersuchungen zu Bevölkerungsdichten, Mobilität und Landnutzungsmustern im späten Jungpaläolithikum*. KSPA 6, Leidorf, Rahden/Westf.

Lindenfors P, Wartel A, and Lind J. 2021. 'Dunbar's number' deconstructed. *Biol. Lett.* 17: 20210158.

Maier, A. 2015. *The Central European Magdalenian. Regional diversity and internal variability*. Vertebrate Paleobiology and Paleoanthropology Series. New York: Springer.

Maier, A. 2019. Das Magdalénien im Elsass, in Südwestdeutschland und in der Schweiz im überregionalen Kontext. In: H. Floss (Hrsg.), *Das Magdalénien im Südwesten Deutschlands, im Elsass und in der Schweiz*. Eine internationale Fachtagung zum 100-jährigen Grabungsjubiläum in der Kleinen Scheuer im Rosenstein (p. 297-322). Tübingen: Kerns.

Maier, A. 2020. On the morphology, chronology, and phylogeny of triangular lithic insets between 20 and 14 calBP in Western and Central Europe. *Anthropologie* LVIII/2-3, 167-190.

Maier, A. and Zimmermann, A. 2017. Populations headed south? The Gravettian from a palaeodemographic point of view. *Antiquity* 91, 573-588.

Maier, A., Lehmkuhl, F., Ludwig, P., Melles, M., Schmidt, I., Shao, Y., Zeeden, Ch. and Zimmermann, A. 2016. Demographic estimates of hunter-gatherers during the Last Glacial Maximum in Europe against the background of palaeoenvironmental data. *Quaternary International* 425, 49-61.

Maier, A., Liebermann, C., and Pfeifer, S. 2020. Beyond the Alps and Tatra Mountains — the 20-14 ka Repopulation of the Northern Mid-latitudes as Inferred from Palimpsests Deciphered with Keys from Western and Central Europe. *Journal of Paleolithic Archaeology* 3(3), 398-452.

Maier, A., Stojakowits, P., Mayr, C., Pfeifer, S., Preusser, F., Zolitschka, B., Anghelinu, M., Bobak, D., Duprat-Oualid, F., Einwögerer, T., Hambach, U., Händel, M., Kaminská, L., Kämpf, L., Łanczont, M., Lehmkuhl, F., Ludwig, P., Magyari, E., Mroczek, P., Nemergut, A., Nerudová, Z., Niţă, L., Polanská, M., Połtowicz-Bobak, M., Rius, D., Römer, W., Simon, U., Škrdla, P., Úivári, G., and Veres, D. 2021b. Cultural evolution and environmental change in Central Europe between 40.000 and 15.000 years ago. *Quaternary International* 581-582, 225-240.

Maier, A., Ludwig, P., Zimmermann, A. and Schmidt, I. (accepted). The sunny side of the Ice Age: Solar insolation as a potential long-term pacemaker for 3 demographic developments in Europe between 43 and 15 ka ago. *PaleoAnthropology*, DOI: 10.13140/RG.2.2.19046.55367.

Milgram, S. 1967. The Small World Problem. *Psychology Today* 1, 60-67.

Neiman, F.D. 1995. Stylistic variation in evolutionary perspective: inferences from decorative diversity and interassemblage distance in Illinois woodland ceramic assemblages. *American Antiquity* 60, 7-36.

Reyes-García, V., D. Zurro, J. Caro and M. Madella. 2017. Small-scale societies and environmental transformations: coevolutionary dynamics. *Ecology and Society* 22(1): 15.

Rindos, D. 1989. Undirected Variation and the Darwinian Explanation of Cultural Change. *Archaeological Method and Theory* 1, 1-45.

Schmidt, I. 2015. *Solutrean Points of the Iberian Peninsula. Tool making and using behaviour of hunter-gatherers during the Last Glacial Maximum*. BAR International Series 2778. BAR: Oxford.

Schmidt, I. and Zimmermann, A. 2019. Population dynamics and socio-spatial organization of the Aurignacian: Scalable quantitative demographic data for western and central Europe. *PLoS ONE* 14(2):e0211562.

Schmidt, I., Hilpert, J., Kretschmer, I., Peters, R., Broich, M., Schiesberg, S., Vogels, O., Wendt, K. P., Zimmermann, A. and Maier, A. 2021. Approaching Prehistoric Demography: Proxies, Scales and Scope of the Cologne Protocol in European contexts. *Philosophical Transactions of the Royal Society B*. 376.

Shennan, S. J. 2000. Population, culture history and the dynamics of culture change. *Current Anthropology*, 41, 811-835.

Shennan, S. J. 2001. Demography and cultural innovation: a model and its implications for the emergence of modern human culture. *Cambridge Archaeological Journal*, 11 (1), 5-16.

Spielmann, K. A. 2002. Feasting, craft specialization, and the ritual mode of production in small-scale societies. *American Anthropologist* 104: 195-207.

Vanhaeren, M. and d'Errico, F. 2006. Aurignacian ethno-linguistic geography of Europe revealed by personal ornaments. *Journal of Archaeological Science* 33, 1105-1128,

Watts, D. J., and Strogatz, S. H. 1998. Collective dynamics of "smallworld" networks. *Nature*, 393 , 440-442.

Weninger, B., Jöris, O., and Danzeglocke, U. 2014. CalPal-2014. Cologne Radiocarbon Calibration & Palaeoclimate Research Package. Version 2016.2.

Zhang, X.Y., Drake, N.A., and Wainwright, J. 2004. *Scaling Issues in Environmental Modelling, Environmental Modelling. Finding Simplicity in Complexity.* Southern Gate: John Wiley & Sons, Ltd. 319-334.

Comment by Robert L. Kelly

I agree with this chapter's position, that archaeology, and especially paleolithic archaeology, is best at understanding the manifestation of human agents at a high level of abstraction, the collective results that produce patterns in archaeological data at a scale of thousands of years. Here we can see changing levels of population and network size/connectivity over a period of nearly 30,000 years in paleolithic Europe. The relationship between population and network size/connectivity, however, is not entirely consistent. On this matter I have two comments.

First, the work here focuses on the role of population, which is certainly important. Low population levels can make network connectivity difficult – it is hard to meet up with members of other residential groups if those groups are spread widely across the landscape. Higher population levels, on the other hand, can obviate the need for long-distance connections or "max-out" an individual's capacity for meeting/knowing others – because there is already too many people to know in the local neighborhood.

What could be added to this picture is some understanding of the variables that condition the extent to which foragers at different time periods needed the social connections with those living in other regions and the extent to which foragers could provide aid to neighbors. This is largely a product of the extent to which two regions (to take a simple case) are in sync climatically or not. Regions that are in sync cannot provide aid in times of need, and do not need aid in good times. Conversely, regions that are not in sync can provide aid, because a resource bloom in one area would be a resource decline in a neighboring one. Necessary to this proposal would be the climate data that permit reconstruction of at least relative differences in the degree of correlation between those regions demarcated here through artifact styles.

Second, the nature of the social connections that create the networks is not discussed here because, in fact, that level of detail is difficult for paleolithic archaeology to achieve. Nonetheless, the nature of the connection among regions may matter. One likely vehicle is marriage (and so the model proposed by Widlok and Henn in this volume might be usefully combined with this chapter's project). Would particular marriage practices adaptively push people to search for mates far afield under different levels of population density? Or would low population density select for cultural practices similar to "walkabouts" so as to create connections (e.g., locate culturally-appropriate spouses)? One might say that the specific cultural practices do not matter – the connections were somehow made (as evidenced by regional similarities in artifact styles). But one wonders if this position is taken for theoretically sound reasons, or only because the specific cultural practices (e.g., second cousin marriage) are impossible to see archaeologically. Agent-based modelling is one way to test different scenarios.

Adding both of these elements into the current project might help understand why population measures alone produce some inconsistencies in the patterns.

Comment by Graeme Warren

In their contribution Andreas Maier and colleagues usefully highlight that what is meant by small-scale can vary considerably, and that the term is used in multiple, and not always consistent senses. Their focus is on population size, density and distribution: using demography as an index of one aspect of scale. This is supported by an analysis of interaction, which is indexed by patterns of regional material culture similarity and difference. Their demographic estimates are based on the Cologne Protocol – which is not presented in detail here but uses site density over time and space as the basis for its calculations. The demographic turn in archaeology of recent years has seen extensive use of either site density or, more commonly, radiocarbon dates as proxies for population. It is important to note that the use of all such proxies is challenging. For the purpose of this comment, however, the demographic reconstructions are accepted as given.

Maier et al. emphasise that their view is of the long term, and that this perspective will show different aspects of scale than those focused on the lifespan of individuals, including potentially highlighting emergent properties of behaviour at the long-term and large-scale. Their reconstructions of Upper Palaeolithic networks and demography in Europe show that aspects of scale relating to demography – the size, distribution and density of population – change significantly and that interaction between these groups also changes.

Significantly, there is no single relationship of demography and interaction: it is not possible to predict changes in scale of interaction simply from changes in population scale.

The period 33-29 ka, for example, sees a 'difference of affluence', with a break down of regional interaction argued have been created by population growth. Population density reduced the need for long distance contact and is argued to lead to the capacity of individuals to manage their networks being transcended. This leads to an increased regionalisation in material culture. This relationship is perhaps unsurprising – at the broadest of levels, for example, population growth in the Mesolithic of Europe has long been argued to see increased regionalisation.

From 29-25ka by contrast, the break-down of regional interaction networks is argued to derive from climate-change driven demographic decline leading to smaller population groups. In such groups it is proposed that long distance contacts were sustained by a comparatively small number of individuals, and that these networks were therefore fragile and susceptible to collapse: in this instance the break down of interaction is a 'difference of loss'.

There is much of interest in these discussions, highlighting the complex relationships that bind demography and interaction. At the same time however, it is worth noting that the movement to the largest of scales and the emphasis on 'faceless collectives' also creates its own explanatory relationships. Variation in insolation is argued to lead to changes in the availability of resources and therefore hunter-gatherer demographic change. This in turn has implications for network maintenance and the scales at which these communities lived. This may well be the case. But by moving to this time scale, climate and the environment has become the only explanatory framework for network change – there is no other data set operative at this scale against which the demographic data can be mapped. This is not to argue that the climate driven explanation is incorrect. But it is important to ask what might have been lost in moving to this scale, and what other perspectives on hunter-gatherer scale and networks – be it small or otherwise – are occluded behind the clouds of climate change. Finding ways of integrating other scales of analysis within the large scale would help provide these different perspectives. In this regard, more detail of the material culture similarities or differences might complement the large scale demography here, allowing some consideration of alternative scales.

Socioecological factors influence hunter-gatherers

Group size, lifetime interactions, and emergent properties of culture

Brian F. Codding, Kasey Cole, and Kurt M. Wilson

Introduction

The size of hunter-gatherer societies varies across multiple scales. In one dimension, individuals may experience a very different number of co-residing others, often referred to as band size.

Consider a Martu woman living in the western Australian desert in the 1940s: she may spend most of her days in a small, tight-knit, highly mobile community of kin that includes her mother and sister co-wives, who cooperate in an all-female division of labor to feed themselves and their kids (Scelza and Bliege Bird 2008; Peterson and Long 1986); yet she also maintains a broad social network linking her to a large number of kin and non-kin who might rarely see one another (Bird et al. 2019).

Contrast this with the life of a Dogrib woman living in the western Canadian subarctic in the 1930s: she may spend the summer fishing with her extended family and many non-kin, the winter processing caribou hunted or trapped by members of her nuclear family plus a few other close kin, and the spring and fall in other task-specific aggregations of varying sizes (Helm 1965, 1968; Andrews and Zoe 1997); her core social network might be built from no fewer, and no more, than these same individuals with whom she co-resides for part of every year on a seasonal round.

These two women may experience very different lives in terms of the number and relatedness of the people with whom they live on a day-to-day basis, but they may interact with nearly the same number of others throughout their lifespan. This is another dimension across which individuals living in hunter-gatherer societies might experience variation: the total number of interactions they experience with a unique other during their lifetime.

Consider an Ache woman living in the neotropics of Paraguay in the 1960s. She might spend her days in a band of about 20 individuals, to most of whom she is unrelated save only her brother whose family also camps with hers (Hill et al. 2011). Nearly every day she packs up her family and their belongings to transport them to a new camp, pausing along the way to forage and assist her current husband in spotting game (Hurtado et al. 1985). From time to time new families join the band, others leave, and throughout her lifetime she may end up interacting with nearly all Ache living in the region (Hill et al. 2014).

This differs from what a Ju'/hoansi woman experiences through her lifetime in the arid Kalahari savanna of Botswana in the 1930s (Lee 1968, 1972, 1979). While she too lives with a similar number of bandmates on an average day, between 15 to 30 (Hill et al. 2011; Wiessner 2014), she experiences remarkable seasonal and inter-annual variation. During the wet season, she might be living in a smaller than average group made up of her immediate family camped in the mongongo groves where she works two or three days a week harvesting and processing nuts (Lee 1968, 1972, 1979). During the dry season, she and her husband might join up with her parents and a few other families to camp near permanent water in a group of about 30 individuals. During the dry season in an extremely dry year, she may be joined by her husband's parents, along with many other families as they congregate at the largest water source in the region in a group that tops 80 people (Lee 1972). Repeated throughout her lifetime, and strengthened by the formal exchange partnerships she cultivates (Wiessner 2002, 2014), she may end up interacting with a large number of people, despite living in a smaller than average band.

These four women may share a similar mode of subsistence — hunting and gathering — but they experience a different number of co-residents day-to-day, and a disparate number of interactions with unique others aggregated over their lifetimes. Moreover, how they interact with others varies, suggesting meaningful differences in the quality, not just the quantity, of interaction spheres. This variation is often averaged over by researchers seeking to find a single number charactaristic of human social organization (e.g., Dunbar 1993). Yet doing so is a missed opportunity to explore, and possibly explain, the variation.

Here we build on recent synthetic analyses (Hill et al. 2011; Hill et al. 2014), and case studies (Bird et al. 2019; Wiessner 2014) summarized in a recent review (Codding et al. nd.) to further explore this variation in hunter-gatherer group size and the degree of interaction, specifically focusing on the influence of ecological, economic, and social factors, and the sociocultural consequences of these socioecological dynamics. We begin with a summary of our theoretical approach, and specific expectations derived from it.

Theory First

Our research strategy follows a tradition established by anthropologists leveraging theory from behavioral ecology to examine variation in hunter-gatherer socioecology (Winterhalder and Smith 1981; Codding and Kramer 2016). A core part of this strategy is an emphasis on individual decision making within a local environmental context, wherein environment is broadly defined to include the natural and social worlds within which an individual resides. The assumption is that individuals will do the best they can to make a living, given the opportunities and constraints afforded by their local context. This assumption is akin to extending the principal of charity (Gauker 1986) or of humanity (Dennett 1987) from interpreting an individual's thoughts and words in the most reasonable way possible, to interpreting their decision making and behavior with the same charitability. Though in this case, we also have a general theory of behavior that provides a function to identify the objective of behavior (*cf.* Gauker 1986: 2-3).

To derive expectations *a priori* about how individuals should behave within a specific context, researchers rely on formal models. These models can examine optimal decisions relative to the distribution and abundance of resources (i.e., resource-contingent decisions; e.g., Charnov 1976), to how many others co-reside in that environment (density-dependent decisions; e.g., Fretwell and Lucas 1969), and to the strategies others employ (frequency-dependent decisions; e.g., Barnard and Sibly 1981). Model predictions are compared quantitatively or qualitatively to observed behavior (or their material correlates in the case of archaeology), with mismatches indicating problems with model specifications, not that the behavior is suboptimal or maladaptive. In other words, following the adaptationist program (Mayr 1983), behaviors are assumed to be near optimal within constraints, with research focused on the simplest explanation possible without being overly reductionist or deterministic.

Out of these decisions, broader patterning emerges, such as sustainable land use (Moritz et al. 2018), divisions of labor (e.g., Codding et al. 2011), hierarchy (e.g., Smith and Codding 2021; Wilson and Codding 2020), ethno-linguistic diversity (Codding and Jones 2013), and other cultural phenomena (e.g., Smith et al. 2017). As the product of individual decisions in a local context, explanation of these sociocultural phenomena requires research focused on individuals and their relations, not on the emergent properties themselves. Thus, the approach provides a way to look at the socioecological foundations of behavior that aggregate to produce patterning expressed as cultural phenomena, including its material correlates studied by archaeologists (O'Connell 1995; Bird and O'Connell 2006; Codding and Bird 2015).

Implementing this general research strategy using cross-cultural evidence, here we propose that three factors derived from three core principals in the ecological and behavioral sciences can be used to predict how and why band size and interaction spheres may vary among hunter-gatherers (Codding et al. nd.). We refer to these as Allee-effects, the maximum sustainable yield, and social dilemmas.

First, we suggest that band size should be structured by returns-to-scale, or Allee effects, associated with focal resources. Within an ideal distribution model framework (Fretwell and Lucas 1969), the decision of any one individual to settle in a resource patch or habitat will depend in part on the number of others who already reside there. For many resources, this relationship takes the form of negative density dependence, wherein everyone does worse with each additional joiner. However, there are also often benefits to aggregation, such that everyone does better up to some threshold as more people co-reside. These Allee-like benefits can arise when aggregations result in increased habitat quality (e.g., Bird et al. 2019), shared defense costs (e.g., Codding et al. 2019), or returns-to-scale in resource acquisition (e.g., Smith 1985). We suggest that those resource acquistion activities with positive density dependence should encourage more individuals to cooperate, thus aggregating to produce larger mean experienced band sizes. In a very simple example, we previously showed that returns-to-scale increase from economies focused on plants, to those focused on aquatic and hunted resources (Codding et al. nd.).

Second, we suggest that the size of an individual's interaction sphere is in part structured by the maximum sustainable yield (MSY) of focal resources. MSY is the highest number of resources that can be taken at that rate in perpetuity. Because foraging depletes resources (e.g., Alvard 1994), individuals may be better off moving camp more or less frequently depending on MSY. This can be modeled following the marginal value theorem (Charnov 1976), which shows that individuals should move and resettle elsewhere following depletion to the point of diminishing returns. Available evidence supports these model predictions (e.g., Aswani 1998; Codding et al. 2016; Venkataraman et al. 2017), though individuals are likely to stay longer than what is sustainable when doing so maximizes their immediate returns, leading to long-term unsustainable outcomes (Alvard 1993; Hardin 1968). Of course, there are also other factors that structure mobility and interaction rates, but all else being equal, individuals should move more frequently if their focal resources have a lower MSY. If individuals reshuffle co-residents when they move, then this should influence interaction rates, though more detailed analysis is required to unpack how strong this association may be relative to other causal factors.

Finally, the strategies that others employ should structure band size and interaction rates through complex dynamics influenced by frequency-dependent decisions (Barnard and Sibly 1981). Specifically, larger band sizes may increase social dilemmas by increasing opportunities to "free ride" on the production of others (e.g., Olson 1965). Producers may find it in their best interest to tolerate an increasing number of non-producers (Blurton Jones 1984, 1987) up to some threshold at which they may decide to "vote with their feet" and leave (Lee 1972; Woodburn 1982) or remain and punish those who do not contribute, the latter being extremely rare (e.g., Gaula 2012; Marlowe 2010). The former may result in high mobility and residential reshuffling that increases interaction rates (e.g, Lewis et al. 2014). In some circumstances, these dynamics may also encourage institutions that restrict who can join groups of producers working in Allee-like (e.g., Codding et al. 2019) or economically intensive (Parker et al. 2019) subsistence activities, which may reduce interactions with outsiders. As such, band size and interaction rates likely feedback on one another depending on how individuals resolve social dilemmas.

This final point highlights how individual decisions might interact with one another to propagate broader social and cultural institutions. Specifically, socioecological factors that incentivize large cooperative groups may encourage "solutions" to collective action problems, while those that induce high lifetime interaction rates may facilitate institutions to manage large social networks. The former may encourage practices and institutions governing group membership, such as territoriality, property rights, and inter-group conflict (Codding et al. 2019; Parker et al. 2019). The latter may encourage practices that help regulate frequent encounters with distantly known others, or that foster long-distance ties with many others, such as ritual practices (e.g., Bird et al. 2019; Hill et al. 2014) or economic partnerships (e.g., Wiessner 2002). If this is supported by further analysis, understanding variation in past socioecology may also help archaeological inquiry ascertain the kinds of social institutions formed by past societies to solve these problems, even if they are otherwise materially invisible.

In summary, building on theory and models from behavioral ecology, we suggest that individual decisions structured by density-dependent, resource-contingent, and frequency-dependent factors will aggregate to produce emergent group-level patterning observed as forager band size and lifetime interactions, which may further feed back to structure decisions that produce emergent social and cultural institutions.

This approach flips some previous perceptions on their head. For example, Barth (1978: 11) notes that "...it is taken for granted that the social institutions of the Andamans simply do not have the capacity to organize large populations" – we would argue it is the other way around: social institutions create capacity in response to population size and interactions, which are in part structured by the local environment. Individuals adapt to their local circumstances; social institutions are the epiphenomenal product of those decisions interacting with and feeding back on one another. In this way, the approach is not environmental determinism, rather, individuals have the agency to determine their actions (Bird and O'Connell 2012). The approach assumes that individuals, having near-to-perfect knowledge of local conditions and expected immediate outcomes, will generally make the best decisions possible out of the available options, which are limited by local conditions. As noted above, this is a charitable assumption. To think otherwise would implicitly assume that individuals lack either the capability or knowledge to do what is best within their local context. Of course, individuals acting in their best interest will not produce outcomes that are best for everyone (e.g., Hardin 1968; Olson 1965), and may result in profoundly negative unintended consequences in the long term (anthropogenic climate change being a prescient example).

Applied to the case at hand, we previously found some support for our predictions (see Codding et al. nd.): mean experienced band size (Hill et al. 2011; Bird et al. 2019) and lifetime interactions with unique others (Hill et al. 2014; Wiessner 2014) seem to co-vary respectively with the primary focus of subsistence and net primary productivity (Binford 2001). Additionally, total population size (and lifespan) will influence the total number of interactions possible (Hill et al. 2014). Finally, a qualitative review of the ethnographic evidence (Codding et al. nd.) also suggests that band size may further influence interaction rates through increased residential reshuffling in response to free-riders (Blurton Jones 1987). Leveraging these predictions, here we explore how these dynamics may structure the range of variation in hunter-gatherer band size and lifetime interactions, and the sociocultural institutions that may emerge from these dynamics.

To explore the empirical record of hunter-gatherer social space relative to our theoretical expectations, here we seek to examine the range of variation in hunter-gatherer band size and lifetime interactions in a sample of 25 societies (see Codding et al. nd.). However, the available evidence required to assess this expectation about lifetime interactions are restricted to only three cases for which data have been quantified: Hadza, Ache (Hill et al. 2014), and San (Wiessner 2014).

Thankfully, this is not an insurmountable problem. Using other information recorded about each society, we can impute estimated lifetime interactants for all societies in this sample. Specifically, here we impute estimates of lifetime interactants for the 22 societies without empirical estimates using a non-parametric imputation method from Stekhoven and Buhlmann (2012) based on the random forest algorithm (Breiman 2001) implemented by Liaw and Wiener (2002). Specifically, we ask the model to predict the 22 missing lifetime interactants based on the three recorded lifetime interactants (lifetime conversations from Hill et al. 2014; interaction sphere from Wiessner 2014) plus documented information on mean experienced band size (Hill et al. 2011; Bird et al. 2019), net primary productivity, the primary source of food (gathered, hunted, or aquatics), and the total population size (Binford 2001).[1]

One other issues stems from the fact that lifetime interaction spheres are estimated differently across studies. For San, Wiessner (2014: Table 2) calculates the number of interactants as the number of adults who co-reside with exchange (*xaro*) partners, with whom individuals will likely interact during aggregation events. She notes this is likely an underestimate of lifetime interaction spheres. For Hadza and Ache, Hill et al. (2014: Table S16) calculate the number of interactants from the results of 16 questions about specific types of dyadic interactions men have with other men, and women have with

1 As imputation error of the *missForest* function varies with the number of variables randomly sampled at each split in a tree (m_{try}) and the number of trees grown in each forest (n_{tree}) (see Stekhoven and Buhlmann 2012), we iterate models through every combination of each parameter from 1 to 100 trees (the default) and 1 to 5 (the max number of variables) to select the combination that minimizes the root mean square error (RMSE) of lifetime interactions, a measure of model fit in the same units as the imputed variable (for the best model, $m_{try} = 5$ and $n_{tree} = 4$). Error rates reported in text are out-of-bag error.

other women, from which they estimate a yearly interaction rate, and a consequent lifetime estimate based on the total population size and lifespan. We take the average number of men's interactants across all arenas and double them to account for interactions with women, which is likely an overestimate for Hadza though may be appropriate for Ache given closer gender parity in interactions (Hill et al. 2014). Together, these provide relative estimates of lifetime interactions that we can use to explore the likely interaction spheres that individuals in different societies might experience.

The model results, shown in Figure 1, predict lifetime interactants for all societies with an error rate (root mean square error [RMSE]) of about 162; meaning the model can on average predict lifetime interactions within plus or minus 162 individuals. This equates to a mean normalized error (NRMSE) of 0.25 (zero being perfect, and one being very poor), which suggests the model is performing reasonably, but not remarkably well. With this in mind, and given the full range of observed variation is between 451 individuals (for Ache) and 870 individuals (for Ju/'hoansi), we can interpret the predicted values as showing meaningful separation at the extremes, but having overlapping errors near the central tendency. In other words, the point estimates in Figure 1 should not be taken as a measure of actual precision, but can be used to explore the possible parameter space characterizing hunter-gatherer group size and interaction rates. Keeping the limitations in mind, we use this graphical result as an opportunity to scope the general bounds and possible range of variation in band size and interaction rates. Here we examine ethnographic cases at the boundaries of this parameter space.

The range of variation in band size is bounded by 5.8 for Hill Pandaram to 81.6 for Dogrib. That of lifetime interactions is bounded by the estimated empirical range from 451 for Ache to 870 for Ju/'hoansi (note, the model cannot impute estimates outside the observed range). Hadza appear near the center of this envelope, which supports Marlowe's (2010) observation about the representativeness of Hadza for warm climate foragers.

Shown in Figure 1, societies living in similar regions and with similar subsistence targets sometimes cluster near one another in this constructed social space, such as Nunamnuit and Dogrib. Other societies living in disparate ecosystems and employing varying subsistence practices find themselves near one another within this social envelope, such as Martu focused on gathering, Apache on hunting, and Ainu on aquatic resources, all of whom cluster toward the central tendency of interaction rates and lower bounds of band size. To explore the socioecological factors that may structure a society's position within

Figure 1: Range of variation in mean experienced band size and interaction sphere for 25 hunter-gatherer societies. Points are color-coded by the main source of subsistence (gathered, aquatic, and hunted resources) with point shape indicating if the interactions value is observed (square) or imputed (circle). Societies mentioned in the text are labeled.

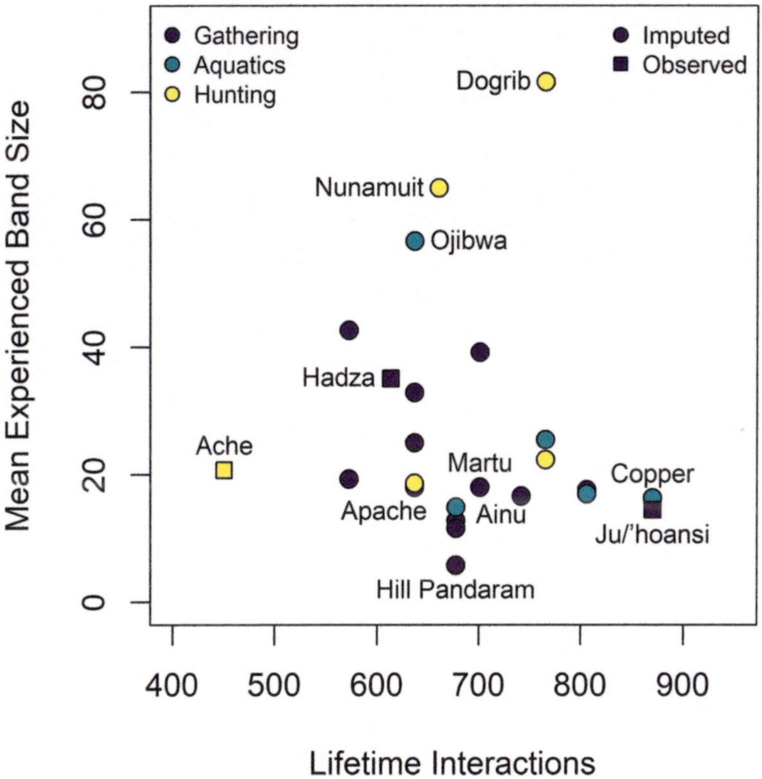

this social space, and the possible sociocultural institutions that emerge from it, we next examine a few cases at the extremes in a bit more detail.

Hill Pandaram

Beginning with the smallest mean recorded bands, an individual living in Hill Pandaram society in the Ghat Mountains of south India may on average live with their nuclear family plus two or three other couples and their children

(Morris 1982). Our analysis suggests this individual might interact with a relatively moderate number of others throughout their life, totalling near 677. Small band size and intermediate lifetime interactions seems to be structured in part by local resources.

Morris (1982) reports a very limited economy of scale, with each individual being responsible for acquiring their own food. Additionally, individuals seem to have intermediate levels of mobility driven by the depletion rate of local resources: "[t]hey remain for a week, and move on when the food supply is exhausted" (Iyer I937: 97; quoted by Morris 1982: 453). These moves every 7-8 days also result in group fissioning down to a single family and fusion up to as many as 20 people on rare occasions. Interestingly, aggregations are not linked with seasonal variation in resource abundance, as is the case for populations with significantly higher interaction rates such as the Ju/'hoansi and Copper Inuit (more below).

These socioecological dynamics may further influence emergent cultural patterning, including their societal emphasis on individual autonomy and self sufficiency, as well as egalitarian roles between women and men, and the lack of institutions to facilitate corporate organization (Morris 1982), all of which may stem in part from such limited returns-to-scale. Further, there are no ritual aggregations, and kinship is "geared to mating and adhesion, not filiation" (Morris 1982: 454), both of which may in part result from the limited need to maintain larger than average social networks.

Dogrib (Tlicho)

At the opposite extreme of mean experienced band size, an individual living in Dogrib (Tlicho) society in the Northwest Territories of Canada may on average co-reside with 82 others (Hill et al. 2011). This average derives from individuals in Marten Lake Dogrib communities residing in about twenty households (Helm 1967; Denham 2010), who regularly aggregate and disaggregate into specific task group formations. These include two to three men who go out trapping, a trapping party of a few families, the aggregation of many families for caribou hunting, and further aggregations for fishing camps, each of which may last from a few days to several weeks (Helm 1965: 378).

Our analysis suggests that the typical person might interact with 765 others through these settlement dynamics, which is high relative to the range of variation and might encourage institutions to facilitate this scale of interactions. Indeed, in discussing kinship, Helm (1965) argues bilateral ties allow flexibility in residence while linking individuals together in a "social chain" (Helm 1965).

While kin ties may be structured to facilitate within-group connectedness, larger – though not necessarily coordinated – seasonally-specific task groups focused on the acquisition of reliable and abundant resources may also incentivize territorial property rights to exclude outsiders (Dyson Hudson and Smith 1978; Codding et al. 2019). Helm (1965: 363) notes some possible support for this, given that the four socio-territorial groups of Arctic Drainage Dene (two Hare, one Slavery, and one Dogrib) who occupied the region at the time of her study acquired resources "in their own region" to which they tied their identity. The lack of even stronger territorial boundaries is not surprising given the spatio-temporal brevity (see Dyson Hudson and Smith 1978) of these larger aggregations, and that the activities undertaken did not require coordinated labor (Helm 1965: 378).

Northern Ache

Northern Ache living in the neo-tropical forests of eastern Paraguay experience relatively small mean experienced band size and the lowest observed lifetime interactions (Hill et al. 2011, 2014). As the hypothetical story in the introduction highlights, Ache live in small, fluid, highly mobile bands with women's tasks focused on plant resources, and moving camp, while men's are focused on hunting mid- to small-sized game (e.g., nine-banded armadillo). Men tend to search for prey independently, but keep close enough so that they may call on other hunters to aid in pursuit. Modeled Ache hunting returns increase with group size (peaking at about 2-7 or more depending on the prey), but observed pursuit group size tends to be smaller than optimal (about 1-5 depending on the prey; McMillan 2000). This may result from tradeoffs between maximizing individual search efficiency by distancing further away from other hunters than is optimal to facilitate cooperation, or from frequency-dependent decisions leading some not to call for help, or others not to join, resulting in failures of coordination (McMillan 2000).

Evidence of high mobility, but low interaction rates seems contrary to our proposed hypothesis. The pattern may result from relative group stability, or the relatively small total population size which limits the possible number of interactants (Hill et al. 2014). The former seems an unlikely explanation, given that individuals are likely to interact with nearly all other Ache through their lifetime, and that they maintain formal ritual relationships which foster connections and ties among highly mobile bands (Hill et al. 2014). As such, their low population size (about 560; compared to 950 for Hadza and 2200 for Ju/'hoansi) is likely the most limiting factor. Further unpacking this finding will require research into the determinants of total demic population size and total territory size, which may also result from the interaction of dynamic socioecological factors (e.g., Parker et al. 2019) to structure the possible number of interactants in a society, and the density of interactions.

Ju/'hoansi (San)

Ju/'hoansi (San, or !Kung) living in the semi-arid Kalahari savanna of Botswana and Namibia have the highest interactions and total population size, though live in smaller than average bands (San mean experienced band size = 14.5 compared to 28.2 for all 32 societies in Hill et al. 2011) and at relatively low population densities (0.16 people per square kilometers; Lee 1968).

The relatively small groups may result in part from the limited economy of scale of focal resources, such as mongongo nuts which are acquired by individual women and men with "one tree to a person" (Lee 1979: 192), though children may help maximize returns by processing nuts back at camp (Blurton Jones et al. 1994). While plants, and especially mongongo nuts, make up the bulk of Ju/'hoansi diet, there are some other resources that encourage cooperation. For example, hunting is more likely to be successful in larger groups (Yellen 1977), men most often pursue game in pairs, and larger cooperative groups formed for game drives (Hitchcock et al. 1996).

High lifetime interactions appear driven by high mobility and seasonal aggregations, which are in part influenced by resource depletion rates and seasonal water availability respectively. Hitchcock et al. (1996: 162) note that in the 1960s "[g]roup aggregation and dispersal patterns were related to the abundance of resources... as resources were depleted in an area, people tended to move out, in part to avoid conflict among group members over the remaining resources." Water availability, and its influence on resource abundance, is one of the major factors that determines aggregation. Lee (1972) reports Ju/'hoansi aggregation and dispersion events in the /gam-/ai/ai areas during the 1920s and 1930s. During the rainy season, a minimum of 11 bands would disperse across the region. During a typical dry season, these bands may aggregate into five groups centered on permanent water sources. During an extreme dry year, the 11 bands would aggregate into as few as two or three locations, resulting in as many as seven bands at a single water hole. One of these locations, "ai/ai was a trading center where people from all points of the compass came to visit, dance, and do *hxaro* trading (and sometimes fight)" (Lee 1972: 138). The need to maintain connections with others for dry season aggregations when resources are scarce are facilitated by social institutions that foster long-term, long-distance relationships of mutual aid known as *hxaro* exchange (Wiessner 2002).

Copper Inuit (Kitlinermiut)

Copper Inuit (Kitlinermiut) and Ju/'hoansi have similar mean empirical band sizes, and our analysis predicts that they might have similarly high lifetime interactions as well, despite the disparate nature of these ecosystems and adaptations. This invites a more detailed review of the factors structuring Copper Inuit band size and interactions. Specifically, we might predict that both societies have different resources which structure similar Allee-like groupings, depleteion-driven mobility, and frequency-dependent fission-fusion. To evaluate this, we examine how focal resources structure band size, interactions, and social dilemmas.

Copper Inuit bands are organized around the nuclear family (Damas 1972). Sometimes nuclear family bands may camp alone, fishing on the winter ice or searching for dispersed caribou outside of the migration season (Jenness 1928; Damas 1972). These bands may aggregate when resource acquisition encourages Allee-like cooperation of one form or another, and may disaggregate when resources are depleted or when seasonal changes alter prey behavior. Three resource acquisition activities are illustrative: intercept caribou hunting, fish trapping, and breathing-hole sealing.

Jenness (1928: 156) notes that during the caribou migration, a collection of family bands would travel to locations where herds are expected, camp for a few days searching for and hunting game, then travel as a group about 10-15 miles (16-24 km) and repeat the process. During one summer, hunters "went on ahead to intercept a herd of caribou they had sighted... This hunt well illustrated the unity of an [Inuit] band. Every individual, man, woman, and child, took part in it....both men and women contributed to the discussion that decided the tactics to be employed." Using coordinated labor and natural topography, they drove 15 caribou, after which they "feasted and idled for two days". This hunt illustrates a relatively extreme example of coordinated labor wherein every member of the extended co-residing band cooperates in the acquisition of a focal resource.

Similar patterns seem to influence group size associated with fish trapping, when families aggregate in the spring and build "three barriers of stones across the streams, leaving narrow openings in the two that lay nearest the sea, and completely closing the highest; whenever a shoal gathered in the upper chamber, they blocked its entrance and stabbed the struggling fish with long, three-pointed spears. Nearly a hundred salmon each weighing from three to twelve pounds lay spread out on the flat boulder around the camp." (Jeness 1928: 237).

While these two activities clearly show how cooperation in resource acquisition may structure co-residence size, perhaps the most well known aggregation events – winter sealing camps – are not a direct product of an economy-of-scale. Smith (1984) notes that winter sealing camps of 50-200 people (reported by Damas 1969) would have 12-50 hunters, while the optimal number of cooperating hunters is around three per breathing hole. He suggests this discrepancy is not due to crowding of prime sealing sites (as individuals would be better off distributing proportionally to the suitability of sealing sites; per Frewell and Lucas 1969), to reduce variance in returns (as harvest rate variance does not decline with group size), or to smooth over acquisition variance via sharing (as storage would be sufficient to accomplish this without invoking collective action problems). Instead, these larger-than-optimal sealing camps may serve another function: information-sharing. Specifically, while larger-than-optimal settlements may accelerate the rate of depletion and exacerbate free-rider problems, these costs may be outweighed by the benefits each individual receives from sharing information on the location and abundance of seals in the area around the camp. He argues this may also explain variation in winter sealing camp size, as the greater the unpredictability, the greater the optimal size of the information-sharing network.

Resource and information sharing may also facilitate connections that strengthen interaction networks, such as enduring social bonds and formal partnerships outside of kin relations (Damas 1972). Though individuals may alter their behavior based on what others are doing, producing frequency-dependent variation that may help resolve social dilemmas. In a contemporary arctic community, Ready (2018) shows that high producing households form reciprocal ties, which indicates that producers bias shares to other producers who are more likely to reciprocate. This may be part of a general trend in which individuals are more likely to cooperate with other cooperators. For example, in public goods games Hadza increase cooperation when they co-reside with more cooperators (Apicella et al. 2012; Smith et al. 2018). If similar dynamics occurred at winter sealing camps, positive assortment of cooperators may serve to reduce collective action problems. Moreover, formal sharing institutions such as seal sharing "flipper associates" may be particularly important to maintain connections among Copper Inuit, who have less defined kin structure than neighboring groups (Damas 1972).

Conclusion

Building on theory and models from behavioral ecology, here we suggest that individual decisions structured by density-dependent (returns-to-scale), resource-contingent (maximum sustianable yield), and frequency-dependent (social dilemmas, specifically free-riders) factors will aggregate to produce emergent patterning in experienced band size, lifetime interactions, and subsequent properties of cultural institutions.

Exploring cases that define the social envelope of mean experienced band size and unique lifetime interactions (Figure 1) reveals some support for this proposed socioecological principle. Mean experienced band size does in part appear influenced by returns-to-scale with focal resources, evidenced well by the extreme cases of Hill Pandaram and Dogrib. The number of lifetime interactions with unique others does seem to vary at least partly as a function of the depletion rate of focal resources, as well as broader patterns of resource availability that lead to seasonal aggregations, such as among Ju/'hoansi and Copper Inuit. Total population size also seems to have a strong influence on lifetime interactions, as shown by Ache. Finally, how these scales of society size interact creates variation in the form of social institutions that emerge to facilitate interactions and solve social dilemmas. In this final dimension, socioecological dynamics seem to aggregate in ways that inform the quality, as well as the quantity, of interactions across these two scales of society size.

While many details remain to be worked out, here we suggest that differences in the scale of hunter-gatherer societies may in part be due to differences in socioecology. Through this exercise, we hope that we have allowed the "scale" of hunter-gatherer societies to emerge not "as an artifact of analysis, but rather [as] an empirical property of the things we study" (Barth 1978: 11). Though our approach does have at least two limitations in this regard.

First, we recognize that generating *a priori* predictions limits initial analysis to those factors which are deemed theoretically important. Second, we recognize that our analytical approach imputes lifetime interactions for a much greater number of societies (n=22) than what is empirically observed (n=3). These may indeed generate artifacts of analysis. However, while these are weaknesses, we also see strengths in these elements. Specifically, understanding variation in the scale of hunter-gatherer social organization requires a framework to distinguish signal from noise. Here we present a conceptual grounding framed in established theory as for why two scales may vary across societies, which allows analysis to focus on the signal. Further, while the specific imputed values are certainly imprecise, and may be wrong altogether, the exercise offers a view into how we might be able to understand patterned variation in in the potential social envelope in of hunter-gatherer social organization.

Together, these help move debate away from simple dichotomies regarding whether hunter-gatherer societies are "large" or "small". As illustrated by the hypothetical vignettes in the introduction of this chapter, there is tremendous variation in the scale at which hunter-gatherer societies operate. There is likely even greater variation beyond the reach of ethnography, which may be revealed through archaeological analyses that leverage theory to circumvent issues with direct ethnographic analogy (see O'Connell et al. 1995). We hope future work can build on this chapter, and this volume overall, to explore the full range of variation in the scale of hunting and gathering societies.

Acknowledgement

We are grateful for the opportunity to present these ideas alongside so many eminent scholars in hunter-gatherer research. We thank the organizers, especially Thomas Widlok and M. Dores Cruz for their support and guidance. We are in debt to Kenneth Blake Vernon for directing us to the principal of charity, and for general comments on an earlier version of this manuscript. We thank members of the University of Utah Archaeological Center and two anonymous reviewers for thoughtful comments on this work.

References

Alvard, Michael S. 1993. Testing the "ecologically noble savage" hypothesis: Interspecific prey choice by Piro hunters of Amazonian Peru. *Human Ecology*, 21(4), 355-387.

Alvard, Michael S. 1994. Conservation by native peoples. *Human Nature*, 5(2), 127.

Andrews, Thomas. D., and John. B. Zoe 1997. The Idaa trail: Archaeology and the Dogrib cultural landscape, Northwest Territories, Canada. *At a crossroads: Archaeology and First Peoples in Canada.* Burnaby, British Columbia: Archaeology Press, Simon Fraser University, 160-177.

Apicella, Coren, Marlowe, Frank, Fowler, James H. and Nicholas A. Christakis. 2012. Social networks and cooperation in hunter-gatherers. *Nature* 481, 497-501. DOI: 10.1038/nature10736.

Aswani, Shankar. 1998. Patterns of marine harvest effort in southwestern New Georgia, Solomon Islands: resource management or optimal foraging?. *Ocean & Coastal Management*, 40(2-3), 207-235.

Barnard, C. J., and Sibly, R. M. 1981. Producers and scroungers: a general model and its application to captive flocks of house sparrows. *Animal behaviour*, 29(2), 543-550.

Barth, Fredrik 1978. Introduction, in *Scale and Social Organization*, edited by Fredrik Barth. Columbia University Press, 9-12.

Binford, Lewis R. 2001. *Constructing Frames of Reference: An Analytical Method for Archaeological Theory Building Using Ethnographic and Environmental Data Sets.* University of California Press, Berkeley.

Bird, Douglas W., and James F. O'Connell 2006. Behavioral ecology and archaeology. *Journal of Archaeological Research*, 14(2), 143-188.

Bird, Douglas W., and James F. O'Connell 2012. Human behavioral ecology. In *Archaeological Theory Today*, 2nd edition, edited by Ian Hodder, pp. 37-61. Polity Press.

Bird, Douglas W., Bleige Bird, Rebecca, Codding, Brian F., and Zeanah, David W. 2019. Variability in the organization and size of hunter-gatherer groups: Foragers do not live in small-scale societies. *Journal of Human Evolution*, 131, 96-108.

Bliege Bird, Rebecca and Brian F. Codding 2021. Promise and peril of ecological and evolutionary modelling using cross-cultural datasets. *Nature Ecology & Evolution*, 1-3. https://doi.org/10.1038/s41559-021-01579-w.

Blurton Jones, Nicholas G. 1984. A selfish origin for human food sharing: tolerated theft. *Ethology and Sociobiology*, 5(1), 1-3.

Blurton Jones, Nicholas G. 1987. Tolerated theft, suggestions about the ecology and evolution of sharing, hoarding and scrounging. *Social Science Information*, 26(1), 31-54.

Blurton Jones, Nicholas G. 2016. *Demography and evolutionary ecology of Hadza hunter-gatherers*. Cambridge, Cambridge University Press.

Blurton Jones, Nicholas G., Kristen Hawkes and Patricia Draper. 1994. Foraging returns of! Kung adults and children: why didn't! Kung children forage? *Journal of Anthropological Research*, 50(3), 217-248.

Breiman, Leo. 2001. Random forests. *Machine Learning*, 45(1), 5-32.

Charnov, Eric L. 1976. Optimal foraging, the marginal value theorem. *Theoretical Population Biology*, 9(2), 129-136.

Codding, Brian F., and Dougals W. Bird. 2015. Behavioral ecology and the future of archaeological science. *Journal of Archaeological Science*, 56, 9-20.

Codding, Brian F., and Karen L. Kramer (eds.). 2016. *Why forage?: hunters and gatherers in the twenty-first century.* Albuquerque, University of New Mexico Press.

Codding, Brian F., and Terry L. Jones. 2013. Environmental productivity predicts migration, demographic, and linguistic patterns in prehistoric California. *Proceedings of the National Academy of Sciences*, 110(36), 14569-14573.

Codding, Brian F., Ashley K. Parker, and Terry L. Jones. 2019. Territorial behavior among Western North American foragers: Allee effects, within group cooperation, and between group conflict. *Quaternary International*, 518, 31-40.

Codding, Brian F., Kasey Cole, and Kurt M. Wilson (nd.) Exploring the ecological foundations of uniquely human social organization.

Codding, Brian F., Rebecca Bliege Bird and Douglas W. Bird. 2011. Provisioning offspring and others: risk–energy trade-offs and gender differences in hunter–gatherer foraging strategies. *Proceedings of the Royal Society B: Biological Sciences*, 278(1717), 2502-2509.

Codding, Brian F., Zeanah, David W., Bliege Bird, Rebecca, Parker, Christopher H., and Bird, Douglas W. 2016. Martu ethnoarchaeology: Foraging ecology and the marginal value of site structure. *Journal of Anthropological Archaeology*, 44, 166-176.

Damas, David. 1969. Environment, history, and Central Eskimo society, in *Ecological Essays*, edited by David Damas, pp. 40-64, Bulletin 230, Ottawa, National Museums of Canada, 40-63.

Damas, David. 1972. Central Eskimo systems of food sharing. *Ethnology*, 11(3), 220-240.

Denham, W. W. 2010 [1978]. Group Compositions in Band Societies (GCBS) Database Manual. https://www.kinsources.net/kidarep/dataset_attachment-/60/216/GCBSManualREADFIRST.pdf.

Dennett, Daniel C. 1987. *The intentional stance*. MIT press.
Dunbar, Robin I. 1993. Coevolution of neocortical size, group size and language in humans. *Behavioral and Brain Sciences*, 16(4), 681-694.
Dyson-Hudson, Rada, and Smith, Eric Alden. 1978. Human territoriality: an ecological reassessment. *American Anthropologist*, 80(1), 21-41.
Fretwell, Stephen D., and H. L. Lucas. 1969. On territorial behavior and other factors influencing habitat distribution in birds. *Acta Biotheoretica* 19(1): 16-36.
Gauker, Christopher. 1986. The principle of charity. *Synthese* 69(1), 1-25.
Guala, Francesco. 2012. Reciprocity: Weak or strong? What punishment experiments do (and do not) demonstrate. *Behavioral and Brain Sciences*, 35(1), 1-15. DOI: 10.1017/S0140525X11000069.
Hardin, Garrett. 1968. The tragedy of the commons: the population problem has no technical solution; it requires a fundamental extension in morality. *Science*, 162(3859), 1243-1248.
Helm, June. 1965. Bilaterality in the socio-territorial organization of the Arctic Drainage Dene. *Ethnology*, 4(4), 361-385.
Helm, June. 1968. The nature of Dogrib socioterritorial groups. In *Man the hunter*, edited by R. B. Lee and I. DeVore, Chicago: Aldine, 118-125.
Hill, Kim R., Brian M. Wood, Jacopo Baggio, Magdalena Hurtado, and Robert T. Boyd. 2014. Hunter-gatherer inter-band interaction rates: Implications for cumulative culture. *PloS One*, 9(7), e102806.
Hill, Kim R., Robert S. Walker, Miran Božičević, James Eder, Thomas Headland, Berry Hewlett, Magdalena Hurtado, Frank Marlowe, Polly Wiessner, and Brian Wood. 2011. Co-residence patterns in hunter-gatherer societies show unique human social structure. *Science*, 331(6022), 1286-1289.
Hill, Kim, Kaplan, Hillard, Hawkes, Kristen, and Hurtado, A. Magdelena. 1985. Men's time allocation to subsistence work among the Ache of eastern Paraguay. *Human Ecology*, 13(1), 29-47.
Hitchcock, Robert K., John E. Yellen, Diane J. Gelburd, Alan J. Osborn and Aron L. Crowell. 1996. Subsistence hunting and resource management among the Ju/'hoansi of Northwestern Botswana. *African Study Monographs*, 17(4), 153-220.
Hurtado, A. Magdelena, Hawkes, Kristen, Hill, Kim, and Kaplan, Hillard. 1985. Female subsistence strategies among Ache hunter-gatherers of eastern Paraguay. *Human Ecology*, 13(1), 1-28.
Jenness, D. 1928. *The people of the twilight*. New York: Macmillan.

Lee, Richard B. 1968. What hunters do for a living, or, how to make out on scarce resources. In *Man the Hunter*, edited by R. B. Lee and I. DeVore, Chicago: Aldine, 30-48.

Lee, Richard B. 1972. !Kung spatial organization: an ecological and historical perspective. *Human Ecology*, 1(2), 125-147.

Lee, Richard B. 1979. *The !Kung San: men, women and work in a foraging society*. Cambridge University Press.

Lewis, Hannah M., Lucio Vinicius, Janis Strods, Ruth Mace, and Andrea B. Migliano. 2014. High mobility explains demand sharing and enforced cooperation in egalitarian hunter-gatherers. *Nature Communications*, 5(1), 1-8.

Liaw, Andy, and Matthew Wiener. 2002. Classification and regression by randomForest. *R news*, 2(3), 18-22.

Marlowe, F. 2010. *The Hadza: hunter-gatherers of Tanzania*. University of California Press.

Mayr, Ernst. 1983. How to carry out the adaptationist program? *The American Naturalist*, 121(3), 324-334.

McMillan, Garnett P. 2001. Ache residential grouping and social foraging. Unpublished PhD. Dissertation, University of New Mexico.

Moritz, Mark, Behnke, Roy, Beitl, Christine M., Bleige Bird, Rebecca, Chiaravalloti, Rafael M., Clark, Julia K., Crabtree, Stefani A., Downey, Sean S., Hamilton, Ian .M., Phang, Sui C. and Scholte, Paul, and Wilson, James A. 2018. Emergent sustainability in open property regimes. *Proceedings of the National Academy of Sciences, 115*(51), 12859-12867.

Morris, B. 1982. Economy, affinity and inter-cultural pressure: notes around Hill Pandaram group structure. *Man* 17(3), 452-461.

O'Connell, James F. 1995. Ethnoarchaeology needs a general theory of behavior. *Journal of Archaeological Research*, 3(3), 205-255.

Olson, Mancur. 1965. *Logic of collective action: Public goods and the theory of groups*. Harvard University Press.

Parker, Ashley K., Parker, Christopher H., and Codding, Brian F. 2019. When to defend? Optimal territoriality across the Numic homeland. *Quaternary International*, 518, 3-10.

Peterson, Nicholas, and J. P. M. Long. 1986. *Australian territorial organization: a band perspective.* Sydney: University of Sydney.

Ready, Elspeth. 2018. Sharing-based social capital associated with harvest production and wealth in the Canadian Arctic. *PloS One*, 13(3), e0193759.

Scelza, Brooke, and Rebecca Bliege Bird. 2008. Group structure and female cooperative networks in Australia's Western Desert. *Human Nature*, 19(3), 231-248.

Smith, Daniel, Philip Schlaepfer, Katie Major, Mark Dyble, Abigail E. Page, James Thompson, Nikhil Chaudhary, Gul Salali, Ruth Mace, Lenora Astere, Marilyn Ngales, Lucio Viniscus, and Andrea Migliano. 2017. Cooperation and the evolution of hunter-gatherer storytelling. *Nature Communications* 8(1): 1-9.

Smith, Eric Alden. 1984. Approaches to Inuit socioecology. *Etudes/Inuit/Studies* 8(1), 65-87.

Smith, Eric Alden. 1985. Inuit foraging groups: Some simple models incorporating conflicts of interest, relatedness, and central-place sharing. *Ethology and Sociobiology*, 6(1), 27-47.

Smith, Eric Alden and Brian F. Codding. 2021. Ecological variation and institutionalized inequality in hunter-gatherer societies. *Proceedings of the National Academy of Sciences*, 118(13).

Smith, Eric Alden and Bruce Winterhalder. 1981. New perspectives on hunter-gatherer socioecology, edited by B. Winterhalder and E.A. Smith, pp. 1-11, *Hunter-gatherer Foraging Strategies: Ethnographic and Archaeological Analyses*. University of Chicago Press.

Stekhoven, Daniel J., and Peter Bühlmann. 2012). MissForest—non-parametric missing value imputation for mixed-type data. *Bioinformatics*, 28(1), 112-118.

Venkataraman, Vivek V., Kraft, Thomas S., Dominy, Nathaniel J., and Endicott, Kirk M. 2017. Hunter-gatherer residential mobility and the marginal value of rainforest patches. *Proceedings of the National Academy of Sciences*, 114(12), 3097-3102.

Wiessner, Pauline W. 2014. Embers of society: Firelight talk among the Ju/'hoansi Bushmen. *Proceedings of the National Academy of Sciences*, 111(39), 14027-14035.

Wiessner, Pauline. 2002. Hunting, healing, and hxaro exchange: A long-term perspective on! Kung (Ju/'hoansi) large-game hunting. *Evolution and Human Behavior*, 23(6), 407-436.

Wilson, Kurt M., and Codding, Brian F. 2020. The Marginal Utility of Inequality. *Human Nature*, 31(4), 361-386.

Winterhalder, Bruce and Eric Alden Smith. 1981. *Hunter-gatherer Foraging Strategies: Ethnographic and Archaeological Analyses*. University of Chicago Press.

Woodburn, James. 1982. Egalitarian societies. *Man* 17, 431-451.
Yellen, John E. 1977. *Archaeological approaches to the present: models for reconstructing the past*. Academic Press.

Comment by Andreas Maier

It is not only in times of social distancing that contacts between people are an essential parameter for human societies. Interpersonal contact is at the base of transmitting information and skills throughout networks and thus fundamental to innovation dynamics, material cultural evolution, and social organization. While the quality of interpersonal contacts certainly plays an important role in this regard (see Damm, this volume), the quantity of interactions with unique others throughout a person's lifetime also holds a high explanatory potential for social phenomena at different spatial and temporal scale levels. Codding et al. offer highly insightful estimates on the quantity of lifetime interactions for different groups of hunters, fishers, and gatherers. They find that the individuals in their case studies interact with 451 to 870 ± 162 unique other individuals throughout their lifetime. While the personal network size can thus be quite different, and irrespective of whether or not these estimates are accurate by plus or minus 200 people, Codding et al. provide a sound idea about the order of magnitude of interpersonal contacts within foraging societies. These numbers are highly valuable also for researchers interested in quantitative analyses of foraging communities of the past. Here it is interesting to note, however, that a low total population size seems to be a limiting factor for total lifetime interaction, as discussed in the example of the Northern Ache. It can also be assumed that the distance between interacting bands plays a role, since it relates to energy investments for movements. If that was the case, it would follow that the number of lifetime contacts for individuals living during the Palaeolithic should be found at the lower end of the inferred spectrum. Against this background, it would be promising to estimate unique lifetime contacts for Upper Palaeolithic hunter-gatherers based on the palaeodemographic data provided by the Cologne Protocol (see Maier et al., this volume). The approach presented by Codding et al. has the advantage that it does not rely on direct observations or counts of interactions over an entire lifetime. To the contrary, observations from only a few well-studied key-communities are transferred to other groups using additional parameters such as mean band size, net primary productivity of the environment, the primary source of food (gathered, hunted, or aquatics), and the total population size. All these additional parameters can be provided for the Upper Palaeolithic in Western and Central Europe. This of course would imply shifting from an emphasis on individual decision making within a local environmental context to averaging contacts per group in a regional environment. Such an upscaling approach appears permissible, given that the transfer functions used by Codding et al. also imply a certain abstraction from purely individual decisions. An attempt in this direction would certainly be worthwhile, since it promises to deliver quantifiable data highly relevant for the study of cultural evolution that is otherwise outside the reach of inferences from the archaeological record. Such estimates on lifetime interactions could then be compared with independent data on regional and large-scale movements of people as can be derived from the transport patterns of lithic raw materials and fossil mollusc shells. Together with information from palaeogenetic studies, they can help refining models about the regional and supra-regional network structure of Palaeolithic hunter-gatherers. Seen in a diachronic perspective, the expected findings may yield explanatory potential for differences in artefact diversity in communities in Western and Central Europe, or for the development of networks during the process of the repopulation of Central Europe after roughly 20,000 years ago.

Scale and Inuit social relations

Ilagiit, parts of each other

Elspeth Ready

Introduction

For this collection, we were asked to reflect on the nature of social relations among hunter-gatherers, with the aim of exploring how social organization relates to group size, how to identify the scale of groups, and how the organization of small groups impacts patterns of decision-making and mobility. Of particular relevance to these questions are two recently published ethnographic works that have touched on questions of scale among hunter-gatherers. The first, Bird-David (2017), emphasizes the unscaleable nature of social relations in "tiny-scale" forager societies, such as the Nayaka with whom she worked in Tamil Nadu. The second, Bird et al. (2019), argues that foragers "do not live in small-scale societies," based on their work with Martu in the Western Desert of Australia.

In this chapter, I reflect on the scale of social relations among contemporary Kangiqsujuarmiut and historical Tarramiut (Inuit), by drawing on the distinction between local organization (local group composition), and social organization (broader interaction networks). I suggest that the concept of pluripresence (Bird-David 2017), which is based in "being-together" with different others, is helpful for understanding Kangiqsujuarmiut social relations on a "tiny-scale": specifically, at the level of extended families within modern settlements. In practice, although kinship plays an organizing role in social interaction in Kangiqsujuaq today, there is limited emphasis on genetic relatedness, while kin and kin-like ties can potentially be activated over large spatial scales and across generations. I argue that this expansive view of kinship is relevant for understanding historical social organization in the Eastern Arctic, because it blurs the boundaries between local "family bands." For the purpose of engaging with the ideas brought forward by the aforementioned ethnographers, I will primarily focus on the fine-grained historical and ethnographic record concerning Tarramiut (Inuit) social organization and mobility. At the end I will briefly return to the more sobering realities of the archaeological record and how the ethnographic evidence reviewed here might contribute useful insight to its interpretation. I begin by considering what aspects of scale are most relevant to this discussion.

What do we mean by "small-scale"?

Both of the contributions mentioned above are in one way or another interested in the scale of "social relations." Bird-David (2017) is primarily concerned with the intensity of relationships produced when social interactions occur at a very small scale, where perhaps the most relevant "scaleable" variables are local group size, spatial proximity, and marriage practices. For instance, she argues that "good marriages," such as sibling exchanges, facilitate the deepening of relationships between interconnected groups of people. Bird-David uses the term "pluripresence" to reference "a particular scalar condition that entails the vivid availability of each member of a community to every other member," and suggests that this condition is not scaleable. A key feature of pluripresence is that perceptions of group membership are based on a sense of social proximity, on "being-together" with a set of distinct persons, rather than by ethnic group identity.

Bird et al. (2019) are also interested in the scale of social relations, again as they relate to group size, geographic extent, and kinship. However, they are focused on how individual-level patterns of interactions scale-up to produce macro-level structures, rather than on how scale might impact how people conceptualize their social worlds. They argue that Martu groups are highly fluid: people come together in groups of different sizes, at different times, for different reasons (e.g., foraging, residing, or social and ritual events), and that there is high turnover in the membership of particular groups. Thus each person can have a unique network of relations that overlaps only partially with the networks of others, and these networks may be quite large and geographically distributed. Enduring groups and distinct boundaries demarcating them are consequently absent, producing a large-scale network that is not nested.

One might simply conclude from this that some foraging societies have "large-scale" interaction networks, and others do not: the mobility of Western Desert peoples may be exceptional, while probably few hunter-gatherers in the past were so circumscribed as the Nayaka. This is undoubtedly true, and we should potentially anticipate at least as wide a range of variation in the archaeological record. But the question of why and how exactly scales of interaction differ, and thus the interest in comparison, remains.

Besides these differences in the research context, these authors are also considering different aspects of scale. I try to clarify this difference by focusing on a distinction highlighted by Bird and colleagues between "local organization (who is with whom at a given time and place)" and "social organization (the expansive and virtual patterns in ties that comprise networks of social interaction)." Bird et al. (2019) are clearly interested in social organization, whereas Bird-David (2017) is focused on local organization (perhaps better referred to as group composition) and, even more specifically, modes of interaction that occur as a result of it.

In what follows I examine local organization and social organization of Tarramiut (Inuit from Hudson's Strait coast of Nunavik) in light of the concepts and ideas brought forward by these authors. I think this is a potentially interesting avenue of thought since, as I will describe below, there are many "shared features" in the groups discussed by the two authors, namely, the centrality of visiting in social life, fluidity in local group composition, each person having a unique network of kin and social relations, and the extension of those networks beyond the local residential group. I consider the possibility that, in the Tarramiut case, "tiny-scale," pluripresent modes of interaction and concepts of identity may actually be compatible with—and potentially even facilitate—fluid social organization on larger temporal and spatial scales.

Site background

Kangiqsujuaq, an Inuit settlement of roughly 800 people on the Hudson Strait in Nunavik, Canada, appears from the sky as a cluster of 100 or so colorful buildings nestled in a steep valley, all within roughly one to two square kilometres (Figure 1). From the shoreline, the village slopes gently upward, with all of the "old" houses in the settlement (mostly built in the 1980s) facing outwards towards the water, providing a view of the spectacular cliffs across the bay. One road out of the settlement leads to the tiny airport on the hill, and then continues on for roughly 10 km to Akulivik, a camping area and the launch point for many hunting activities. Beyond that, there are no roads: the nearest settlement is Quaqtaq, home to roughly 400 people, 140 km away as the crow flies. Besides hunting and camping trips within a day's travel of the settlement, when people leave the settlement they mostly fly by airplane, to visit friends and family in other villages, to attend meetings, or to go to doctor's appointments. Meetings and medical appointments often take place in the regional centre, Kuujjuaq, which has a population of roughly 2700. It takes about half a day to reach Kuujjuaq by plane if the weather cooperates. People occasionally travel to the "South"—usually Montreal—for meetings and specialist appointments that cannot take place in Kuujjuaq, or simply to go on holiday. Most imported supplies for the settlement come via one or two sea lifts that arrive during the summer months, while small quantities of perishable goods, like fresh fruits and vegetables, are flown in roughly once a week.

The settlement today seems very "new": since 2013, a huge cooperative store, two administrative buildings, and nearly 100 new housing units have been constructed (increasing from roughly 150 to 250) in order to address crowded housing conditions. The new homes thus represent roughly 40% of all housing units in the village today, meaning that the physical size of the settlement has greatly increased in the past decade, while the composition of households has scaled down considerably, to smaller divisions of extended families.

Despite the recent building boom, the age of the settlement is marked by a few old buildings, like the Catholic mission, which dates back to 1936, and by archaeological remnants of the Révillon Frères trading post established in 1910. In the early decades of the twentieth century, Inuit congregated at these sites seasonally for trade and social events, with some settling more permanently. Sedentarization was accelerated after the Second World War when the Canadian federal government required Inuit families to settle in villages and to send their children to school in order to receive government assistance payments. As suggested above, the settlement today has most modern amenities, but Kangiqsujuarmiut still must travel to larger centres for doctor's appointments and higher education.

Like many other settlements in the Canadian Arctic, the specific location of the modern settlement was guided by decisions made by former traders and missionaries rather than by Inuit choice (Damas 2002), but unlike many other settlements, Kangiqsujuaq is in an excellent location for hunting a wide range of land and sea animals. Inhabitants of the region have long resided in the area: an archaeological site with pre-Inuit subterranean houses is located only a couple of kilometres from the modern settlement. Today, subsistence activities continue to provide food, are a primary focus of recreational activity, form an important basis of cultural identity and pride, and act as a cementer of familial and other social bonds (Ready and Power 2018; Ready 2019).

Figure 1: Kangiqsujuaq as seen from the hill to the southwest of the village, in 2011. Wakeham Bay and the small harbour lie to the left of the picture, the body of water in the background is a small lake, Tasialuk. More recently, houses have been built right out to the edge of the lake.

Photo by the author.

I have been conducting research in Kangiqsujuaq regularly since 2011, when I first travelled there as part of an archaeology team from the Avataq Cultural Institute, to investigate two local sites of interest: the aforementioned sodhouses, and an historical walrus butchery site at Aivurtuuq (literally, the place where there are walruses). The latter is roughly 30 km from today's settlement but still also a location regularly visited for hunting and camping. I returned again in summer 2012 to study Inuktitut, the local language, and eventually spent an entire year in the village in 2013-2014. Since then I have continued to make regular visits, lasting from a couple weeks to a couple months.

Except for the first two summers I spent there, when in Kangiqsujuaq I have had the extraordinary privilege of living with a family. This family includes not only the members of the specific household that I have lived with, but the broader extended family to which they belong as well—numbering in the dozens, albeit with different intensities of interaction and closeness. Outside of my "working" hours spent conducting interviews or doing other research activities in the community, my social life in Kangiqsujuaq is primarily centered within this family. I cook and eat meals with them, go on weekend hunting and fishing trips, spend evenings sitting around the table doing craftwork or playing cards, watch tv, go to church, and attend community events in their company.

Having conducted surveys with 75% of households in the community, as well as a longitudinal series of interviews with households from a number of different family groups, I cannot think of any particular reason that the overall set of people I interacted with socially in the community would be unrepresentative (although some persons or households within that larger set might have some atypical characteristics). Over the past several decades, other families in Kangiqsujuaq have also hosted anthropologists, and ethnographers working elsewhere in the Canadian Arctic have similar experiences of becoming embedded within extended families (Briggs 1970 being a classic example). It is important to emphasize how my experience in Kangiqsujuaq has been profoundly shaped by my age, gender, and relationship status: much of my social interaction has involved other women of similar age, many of whom were also single (though most have children; see Ready 2018 on household composition in Kangiqsujuaq). Thus, although I participated in family gatherings and camping trips with both men and women of all ages, and although I have conducted many interviews with men of different ages, I obviously know much less about social relations among men. The arguments I present below draw on my (a non-Inuit anthropologist's) situated experience in the community, as well as my understanding of interviews I conducted (which were not collected with this paper in mind) and of evidence from historical and contemporary Inuit studies.

I now turn to a discussion of different aspects of scale in Inuit social relations, focused on drawing out connections to the ideas brought forth by Bird-David and by Bird and colleagues. I will first focus on the "quality of quantity" in social relations in Kangiqsujuaq today, meaning the characteristics of relationships among persons, rather than on the actual size or composition of groups. Then, I will examine the structure of Inuit social organization on a broader scale, focusing on historical reconstructions and reports of social organization and settlement patterns in the Kangiqsujuaq region. Finally, in the discussion I attempt to synthesize this evidence, reflecting on the potential implications for archaeology.

The scale of Inuit social relations

Concepts of kin

Inuktitut words for kin reflect the interdependent nature of persons: the base *ila-* "designates any group that is solidary for a short or long period of time" (Graburn 1969: 64), meaning that "both immediate and extended family are primarily understood as composite wholes divided into individual components" (Dorais 2020: 104). Thus, *ilakka*, my relatives, literally means my "co-parts," and *ilagiit*, family, means those who are component parts for each other, in the sense that they are like the different ingredients that constitute a cake, rather than like slices of a cake (Dorais 2020). As implied by Graburn's definition, however, the "group" implicitly referenced by these "component parts" is not necessarily fixed.

The reality of kinship in Kangiqsujuaq is that even siblings may have quite distinct set of relatives from each other. This can be partly attributed to the prevalence of customary adoption and the fact that young people may "test out" partners in their teenage/early adult years before settling on a more permanent relationship (see Collings 2014). For instance, a young woman's first child might be adopted, often by a grandmother or aunt, leading extended family members to be connected as kin in multiple ways (e.g., two girls might be biological cousins but also adopted aunt and niece). People frequently maintain relationships with biological, step-, and adoptive parents, full, half- and adopted siblings, and their families. Though some published definitions for the term (Schneider 1985; Dorais 2020) suggest that *ilagiit* references blood relatives; Saladin d'Anglure (1967) explicitly includes affines, adoptive relations, and step-relatives within its scope for Kangiqsujuarmiut. My impression is that local practice reflects the latter, more expansive, concept—that is, people's ideas of who is kin is not restricted to biological relatives.

Beyond genetic, adoptive, and affinal relations, name-soul, or *sauniq*, relations are an extremely important way in which close ties between persons are affirmed or (re)activated. A *sauniq* is named after another person, usually a deceased relative, and is considered to share their name-soul and therefore some personality traits. Thus a child named after her great-grandmother might be called "mother" by her grandmother. While I cannot treat the subject in detail here, Jessen Williamson (2011), Flora (2019), Trott (2005) and Dorais (2020) provide longer treatments of the name-soul concept and naming practices. An important consequence of name-soul relations is that they provide a way to perpetuate close ties among people across generations.

Balikci (1964) conceptualizes the Inuit *ilagiit* as an ego-centered network within which people have unique sets of relatives, as well as some leeway to decide which sets of relations (including affines and other "social" kin) they wish to associate with most closely. However, Trott (2005) has critiqued Balikci's model, suggesting that Inuit view the *ilagiit* as concentric, starting with the household as nucleus and expanding outward, with cross-cutting links created by naming practices (Saladin d'Anglure 1967 describes a similar model). However, even with such an emic conceptual structure, persons are still uniquely situated due to their particular set of kin and set of names. Consequently I think the concept of an ego-centric network is a useful heuristic for this fact (at least etically). Each person has a different set of potential relationships that may be activated in different social contexts, and at different times, but these ties do need to be activated and maintained, whether that be through food sharing (Bodenhorn 2000; Trott 2005), or through time spent together (see below).

Modes of "being together"

Kin, broadly defined, tend to be an important focus of social interaction in Kangiqsujuaq today. A great deal of social activity is organized by and centered around strong bonds between women who are relatives. Of course, as discussed earlier, this impression reflects my gendered experience there, but I think that it is important to note the importance of bonds between women, given that much of the historical literature on Inuit focuses on local group organization being centered around male ties (such as father-son or brother-brother partnerships). I am not sure to what extent my observations might represent recent change or, perhaps more likely, that close ties among women were simply less of a focus of attention in the classic literature.

There are many kinds of social activities in Kangiqsujuaq (from family birthday parties to village feasts and sporting events), but visiting someone in their home, *pulaartuq*, is the most common. *Pulaartuq* does not require advance planning, nor does it necessarily even require conversation or conjoint activity. A visitor can simply drop-by, and hosts are not necessarily expected to stop going about their business, if they have things to do. A visitor may simply sit on the couch for a while. My understanding is that visiting means that you were thinking of the other person and wanted to see them; it is this action in itself (and not the specifics of the conversation or activity undertaken) that is considered meaningful.

For some of the women I worked with, safety and comfort were often found in the company of others. Elders I interviewed about health and well-being explained that particularly when people are going through a difficult time (such as illness or grieving), that person should not be left alone and their family will ensure that there is someone there to sleep with them in the house, or even in the same room. Closeness, both physical and emotional, among friends and relations is reflected in an extreme attention to detail in people's behaviors, attitudes, and habits; such that one is often expected to anticipate other people's needs (e.g., being hungry, or being cold) without being asked. This anticipation of the needs of others also works in reverse; statements may be intended to prompt the addressee to reflect on the other person's concerns and thereby deduce the existence of a problem, without it ever being verbally acknowledged. For instance, the remark that "the door was open" may be an admonition to pay more attention and close it correctly; and the statement "I don't have any gas" is quite likely a request for help to pay for hunting supplies. A lot of help between people is therefore given or received without it ever being openly requested; indeed, explicitly asking for help without appropriate cues imposes an obligation on the other to help, and may be viewed as an imposition on that person's autonomy. In contrast, indirect requests are more easily ignored, if the other person cannot or does not want to assist (Collings 2014). Others have argued that this orientation towards the feelings and needs of others is an important component of Inuit worldview (Briggs 1970; Nagy 2006; Collings et al. 2017).

The emergence of peer groups as a focus of interaction in Inuit settlements has received considerable attention from ethnographers (e.g., Rasing 2017), but I found that even in social events organized in what might appear to be peer groups (e.g., five or six women in their 20s and 30s gathering to cook and play cards), most of the persons involved were related in one way or another, whether genetically, through adoption, affinally, or through namesakes, and often through more than one of these ways. Part of the reason for this may be statistical (a substantial portion of age-peers may be relatives), but I think that closeness between relatives in older generations (again, perhaps especially between women) channels social interaction, leading their children to become habitual playmates, and often, lifelong friends. This closeness reaffirms that they are kin. In contrast, when relatives do not regularly interact through sharing or visiting, this weakens ties, and people may express concern about this by saying that people are not behaving "like family" anymore. To put it differently, the meaningful part of being "kin" is in the fulfillment of social relationships—expectations of mutual aid and interaction—more than in the fact of genetic relatedness.

However, fulfilling expectations of mutual aid and interaction can consume substantial time, energy, and resources. Today, it is impossible for everyone to fulfill these kinds of obligations with everyone else in the settlement, or even with all of their relatives. As one interviewee mentioned: "I know we're getting bigger, more populated, so it's hard to give away meat all the time when there's a lot of family on his side and my side." Groups of relatives in the past surely also grew (via birth, marriage, and other ways of making kin), and the sets of relationships emphasized consequently changed, but the visibility of this process and the scale of population growth have increased in the modern settlement. Although ties that have become distant can potentially be re-activated and mended, family are the people who help, who visit, and who share. People's spheres of social interaction do not encompass the entirety of the settlement but are concentrated on strong social connections based on biographies of interaction that are often structured by kinship.

Beyond the settlement, and making new family

Though settlements may contain multiple kin-focused communities of social interaction, these social groups are not completely bounded within settlements, nor are they fixed in their composition. Despite the fact that travel between communities today is almost exclusively by air, visiting friends and relatives beyond the local settlement remains extremely important and is facilitated by subsidy programs that have the explicit purpose to "preserve the integrity of the culture and lifestyle" of the region (quote from the Air Inuit website). People, especially young people and the elderly, will go on trips to stay with family in other villages, sometimes for weeks (or months!) at a time. There are many reasons for such travel: an escape from tensions or problems at home; accessing resources not available in a person's home community (e.g., beluga); a change of scenery or a desire to reconnect with other family members; even prospecting for romantic partnerships in a place where fewer people are close relatives.

Staying with someone is perhaps the easiest route to establish social ties in a community. One must of course first somehow secure an invitation—but my impression is that this is not too difficult, even for an anthropologist with only an indirect, non-kin connection (e.g., a friend's friend in the neighboring village). Inuit also make friends with people in other communities in various other contexts; for example, bible camps, training seminars, or hockey tournaments often organize local hosts for participants. These connections may be reactivated later for other kinds of visits.

A common question used in a first conversation with a new arrival in Kangiqsujuaq is "who do you stay with?" In the past, it was generally considered rude to ask people who they were; it was expected that one could figure this out through pathways of mutual connections (Dorais 2020). Although today some Inuit (especially children) are less shy to ask who you are ("*Kinauvit?*"), the more subtle question "who do you stay with" allows people to figure out "with whom" a person belongs in the community and how they might orient themselves to them socially (for instance, whether it would be appropriate to go *pulaartuq* at their residence). My association with a particular family—in stark distinction to most visiting *qallunaat* (white people) who live alone or stay at the hotel—was critical in helping me establish a social circle in the village.

Not surprisingly, kinship metaphors are often used to denote close relationships with non-kin. I was on occasion jokingly referred to as a *tiguaq* (adopted) child of the family I lived with—a designation which also humourously emphasized my cultural incompetence, since I am about the same age as my host. On several occasions—usually in the context of complaints about the irritating or tactless behaviors of other *qallunaat*—my friends and "family" in Kangiqsujuaq took care to mention that they didn't think of me as a "researcher," but rather as "just Elisapie" (Elspeth is a Scottish form of Elisabeth, and Elisapie the Inuktitut form, which quickly became my nickname). Thus ethnic or other indicators of "out-group" identity can be effaced in order to emphasize closeness to persons with whom they have developed strong social relationships. In this case the emphasis is placed on distinct (positive) attributes of the person, their name (ideally shared with someone), and characteristics of that person's social relationship with others (e.g., like a daughter or a brother). Romantic relationships are of course another way to bring someone into the family.

New, unrelated, persons can therefore be folded into local "tiny-scale communities," if they have dedicated the time and energy into "being-with" others. Shared names can facilitate this process. Nevertheless, difficulties related to being an "outsider" (including Inuit from other settlements) can emerge, particularly if there is conflict over resources. In such cases, other kinds of distinctions or criteria for group membership may be mobilized. For example, Inuit who have "married-in" to Kangiqsujuaq have smaller ego-centred family networks in the community (being only associated with their spouse's family), and consequently may have less access to food through sharing. Such individuals will sometimes lament that sharing of country foods (particularly those in limited supply, like beluga) is too focused within *ilagiit* and feel that they should be distributed more widely.

To conclude this section, Inuit settlements today have been shaped by colonial policy, land-claims settlements, and other modern institutions that have restricted residential mobility and drawn distinct boundaries on "groups" at different levels, from settlements (e.g., Kangiqsujuaq), regions under specific land claims (e.g., Nunavik), to all Inuit territories (e.g., Inuit Nunangat). Pan-Inuit identity today is very strong (Morin and d'Anglure 1995; Mitchell 1996), and serves to coordinate cooperative action and mutual aid in a variety of contexts, including online. Although modern settlements in Nunavik today range from roughly 200 to nearly 3000 people, I hope to have demonstrated that, when viewed from the inside, the settlements are already "large-scale," consisting of multiple, smaller close-knit communities (Collings 2011), within which patterns of interaction have some of the "unscaleable" qualities described by Bird-David, most notably, that kin and kin-like ties are activated through close social interaction or economic interdependence (Bodenhorn 2000).

Historical group composition and social organization

Saladin d'Anglure (1967) reconstructs settlement patterns for the Hudson Strait region based on interviews conducted with elders in the 1960s and reports from early explorers of the region. The residents of the south coast of the Hudson Strait from Hopes Advance Bay to Cape Wolstenholme were referred to as *Tarramiut*; other regional groups in what is now Nunavik were the *Qikirtamiut*, on the islands in eastern Hudson's Bay, *Itivimiut* on the eastern coast of Hudson's Bay, and *Siqinirmiut* on Ungava Bay. These groups were distinguished by some dialect differences, which still occur today. Graburn (1969: 35), on the basis of historical sources and his own interviews with informants in the Tarramiut region, noted that "these groupings have indefinite boundaries and are of little significance in differentiating the major cultural features of the area." Saladin d'Anglure provides a total population estimate of 250 for the Tarramiut around 1900, but suggests their earlier population may have been somewhat larger. There were six territories occupied by Tarramiut, most with 15-50 residents, with the Kangiqsujuaq area being the most populated, with 120-140 people.

Seasonal variation in Inuit settlement in the Tarramiut region occurred, though not to the extreme described in Mauss and Beuchat's (1904) influential treatise. In the late 1800s/early 1900s, summer camps were reported in the range of up to 40 people. Because of the seasonality of early explorers' activities, few early reports on winter camps are available. The elders Saladin d'Anglure worked with could not recall large winter villages, but suggested that winter camps were slightly larger than summer camps. Graburn (1969) suggested winter camps ranging from 15 to 60 people. Winter camps in the Tarramiut and Itivimiut region may have been smaller than in other regions due to the local practice of seal-hunting in open water or at the ice floe edge during winter, which can be done alone (as opposed to breathing-hole hunting, which is more productive with a group of hunters; Balikci 1964; Saladin d'Anglure 1967). Stupart (1886) and Payne (1889) described a winter village in the Kangiqsujuaq region that reached 150 inhabitants, although Saladin d'Anglure suggests this large grouping may have been a result of the presence of a research station in the area.

Saladin d'Anglure (1967) describes a nested, multilevel structure for Tarramiut social organization, starting with nuclear families, which were nested within domestic groups that themselves were part of family bands of 20-30 persons. These bands tended to be organized around particular family leaders (referred to as *–kkut*, e.g., *Jaanikkut* meaning "with Jaani/John," a mode of reference still used today). However, he also noted considerable flexibility in the composition of groups at all these levels. Indeed, while this description might provide an accurate "snapshot" of group composition at certain points in the annual cycle, there are several reasons that this description should not be "scaled up" to describe social organization over longer periods of time.

First, frequent remarriage suggests that the reconfiguration of domestic groups ("households") was not rare. "Not too distant" exogamy probably best describes traditional marriage practices in the region: Saladin d'Anglure (1967) suggests that non-relatives were preferred as marriage partners, while Graburn (1969) suggests that marriage "was usually a compromise achieved by marrying distant kinsmen or close nonkin." Both authors agree, however, that conjugal relationships "were relatively fragile and ruptures frequent" (Saladin d'Anglure 1967: 155), and people often re-married multiple times due to the death of a spouse.

Second, as argued earlier, for Inuit "family" is a dynamic and expansive category, meaning that "family bands" were not necessarily composed of same set of people through time. There may have been latitude for choice in residential location while still remaining with family—including, of course, "family" created through adoption, marriage, spousal-exchange, and name-soul relations. Adoption practices were widespread in the past; meaning that like today, many people within a family band would have been likely to have a distinct set of relations extending outside of the current residential group. Trott (2005) suggests that naming practices may have also facilitated the exchange of people. For instance, a child might "belong" in the place where their name-soul previously resided and consequently be adopted to someone at that location. A related observation (see below) is that Inuit appear to have been eager to gather together in larger groups whenever the conditions provided an opportunity to do so (Damas 2002), providing opportunities for groups to reconfigure. I suspect that visiting practices on a smaller-scale (e.g., a person going to stay with other relatives for some time, as they do today) also occurred in the past.

Finally, people's range of movement, or of intermarriage for that matter, was also far from restricted even to within sub-regional groups. As suggested earlier, sub-regional group names like "Tarramiut" should primarily be considered to be geographic designations rather than indicators of distinct groups in social, cultural, or reproductive senses. For instance, Graburn (1969) suggests that during the 19th century, Tarramiut regularly ventured all the way to Kuujjuaq for trade, and that there were yearly meetings of people from throughout Nunavik during inland summer caribou hunts. He also notes friendly relationships between Tarramiut and South Baffin Islanders, and that travel across the Hudson Strait by *umiaq* (skin boats that could hold 20-30 people) occurred regularly. Many people had relatives on the other side of the Strait (and still do today). The distance as the crow flies from the coast near Kangiqsujuaq to the coast of Baffin Island is roughly 145 kilometres, although there is a large island at around the 120 kilometre mark. Distances travelled could be even more extreme: in the early 20th century, a hunter from Ulukhaktok travelled all the way from Victoria Island in the Inuvialuit region to Baffin Island and back again by dogsled (Collings, personal communication). The point is that people clearly interacted with their neighbours and even with people from hundreds (sometimes thousands!) of kilometres away, for trade, for marriage, to acquire rare or highly aggregated resources (e.g., timber, caribou herds, walrus), or even just for the adventure.

Actual residential group composition would therefore likely have had considerable turnover across seasons or years, while still fitting the overall scheme outlined by Saladin d'Anglure. I do not know of any sources that could test these propositions about potential turnover in group composition with historical data from Nunavik, but Damas' historical reconstructions provide suggestive evidence for the Central Arctic: "Although there was much fluidity of personnel among the bands within the major regions, it appears that 60 to 70 per cent of the members of one winter's sealing aggregation band assemblage would return the following winter so that a core of members remained from year to year [among Iglulik, Netsilik, and Copper Inuit]" (Damas 1969: 224). In that region, with winter camps averaging about 100 persons according to Damas, that means 30-40 winter co-residents would turn over from year-to-year. Damas (1969: 130) further notes that "non-kinship features, for example [formalized sharing] partnerships, among the Central [Inuit] may be of equal or greater importance than kinship features in the social structure of some hunting bands." Because such relationships may involve unrelated or distantly-related persons, Damas uses the term "non-kinship" to describe them. In practice, however, I would argue that these are a kind of quasi-kin because kinship relations are substantiated through relationships of mutual aid.

Mixing and interdependence—both within and between regional groups—fits with evidence that conflict between Inuit groups in the Eastern Arctic was not the norm. Most sources, including oral histories, agree that Eastern Arctic Inuit were wary of strangers; but also recount the methods employed to determine that a stranger was not actually a stranger, protocols for signaling friendly intent, and even welcome celebrations for new arrivals (Bennett and Rowley 2004). The point here is not that all relations were harmonious: conflict occurred, within and between local groups, with Cree or Dene in certain regions, and with European visitors and colonizers. But unlike in the Western Arctic, organized warfare appears to have been absent among Inuit in the Eastern Arctic, which Darwent and Darwent (2014: 183) relate to the absence of higher-level political organization ("nations," Burch 1998), resulting from the need "to adapt to greater distances among reliable and sufficient food resources" in the East. Periods of resource scarcity appear to have been important in local, small-scale conflicts (Saladin d'Anglure

1967; Graburn 1969; Fossett 2001; Darwent and Darwent 2014). Damas (1969) notes that the Kiluhikturmiut (Bathurst Inlet) had low levels of relatedness compared to other bands, which may have reflected an assembly process in a depopulated area as a result of migration and/or subsequent to an epidemic. Conflicts with Europeans may have been linked to the fact that Europeans were short-term visitors with novel resources. Additionally, the accuracy of some of the historical evidence for Inuit/non-Inuit conflict in the Eastern Arctic may be questionable (see for e.g., Csonka 1993, 1999).

Discussion

Patterns of interaction within Inuit settlements today are highly structured by kinship. As among the Nayaka (Bird-David 2017), Inuit social lives are focused on interaction with persons who are kin often in multiple (not only biological) ways, including through naming practices, as well as by being neighbours, classmates, co-workers, and so on. Kangiqsujuarmiut place strong emphasis on the importance of being-together and helping each other in everyday life, and this appears to be an important criteria for belonging at the "tiny-scale."

At the same, the historical Tarramiut case adds support to Bird and colleagues' (2019) warning that we should not conflate descriptions of group composition with the scale of individual mobility and social networks. Focusing on the former may lead us to underestimate the latter, giving the impression that social worlds were smaller than they actually were. For Tarramiut, even if the size of most local groups may have been relatively small at any one point in time, the search for marriage partners, the avoidance of conflict, the pursuit of food, the desire to obtain trade goods, and the pleasure of visiting all drove people to move around—over extremely long distances in some cases—and interact with others. Thus, a social life centered around intense and intimate relationships with kin and quasi-kin does not mean that networks of interaction were small or fragmented.

Where there is considerable turnover in group composition over time, then mechanisms for the incorporation of new persons into local groups, as described by Bird-David (2017), seem essential. Indeed, the historical evidence suggests relatively rare inter-group conflict in the Eastern Arctic, and a variety of mechanisms for successfully dealing with newcomers. In Kangiqsujuaq today, tiny-scale communities without distinct boundaries, where membership is based on "being-together," exist in parallel with a shared ethnic identity, in a settlement that is relatively large.

One aim of this collection is to reflect on how we can "scale" insights from the ethnographic to the archaeological record. Archaeologists need theory adequate to the available material record, but the whittling of rich theories of human behavior down to bare bones (literally, in the case of zooarchaeology) is a process that inevitably makes ethnographers wince. Kelly, in his contribution to this volume, outlines differences in scale and type between ethnographic and archaeological theory and data. Here I leave the difficult problem of aggregating the predictions of fine-grained models of behavior to archaeological scales aside, and focus instead on history and ethnography as a way to interrogate the assumptions of behavioral models that are already in use by archaeologists (e.g., foraging theory, models of settlement and mobility such as the ideal-free distribution). I wish to make two points based on the evidence I have presented here.

The first point is that conceptualizing foraging groups or bands as enduring clusters of (the same) people may hide how inter-group relations enable reproduction and resource acquisition over landscapes and through time. In many (of course, not all) ethnographically-observed cases, forager camps are ephemeral constellations of people that come together at a moment in time for various reasons. These temporary clusters are parts of much larger interaction networks, as many authors have recently noted (Hill et al. 2014; Blurton Jones 2016; Bird et al. 2019; Boyd and Richerson 2020). Persons within local groups have distinct networks of kinship and other social relations, both within and beyond current residential groups. These networks direct their movements through the seasonal cycle and through their lifetimes, as their set of ties changes, as well as in response to changes in resource availability.

When kin and kin-like relations are geographically dispersed, neighbouring groups will often comprise kin, affines, or potential marriage partners with whom one's fitness is interdependent. And when groups frequently dissolve and reform in new configurations, last year's neighbour is this year's hunting partner. Characteristics of the resource base and population density will likely be important factors in shaping the benefits of interdependence at different spatial scales. For instance, I have suggested that there may have been a considerable turnover within family bands in the Eastern Arctic, but these groups were undoubtedly more stable than what Bird et al. (2019) have suggested for Martu. Although I have not been able to fully explore the differences here, the variation between and within Eastern, Central, and Western Arctic Inuit are highly informative in this respect (Burch 1998; Damas 1969).

This bring me to my second point, which is that competition (or even benign non-interaction) among small foraging bands may often not be the appropriate framework for thinking about the mobility and resource use strategies of foragers. Recently, Boyd and Richerson (2020) have argued that, "contrary to the conventional wisdom, people in late Pleistocene and Holocene hunter-gatherer societies regularly cooperated in large groups to produce collective goods." They describe abundant evidence for communal foraging activities (e.g., caribou and bison drives, construction of large fish weirs) that would have required multi-band cooperation, as well as evidence that war among hunter-gatherers tended to occur between ethno-linguistic groups (which they estimate as being 500 to a few thousand people), rather than among smaller local groups. They use this evidence as support for the hypothesis that inter-group cooperation is a fundamental component of human adaptation.

My contribution is to suggest that in cases where turnover in group composition is high and where "kin" are dispersed in neighbouring groups, the basis for "in-group" identity is not likely to be found within the local residential group, but rather at a much larger spatial and demographic scale. The alternative possibility to band-level group identity that I have discussed here is that people's perceptions of their social worlds were focused on expansive kinship (and kinship-like) networks. The ethnographic literature is full of examples of social mechanisms, like naming practices, spousal exchange, or gift exchange, that facilitate the maintenance and extension of trusting kinship and kin-like ties over space, and even after death (Wiessner 2002; Bliege Bird et al. 2018). We can potentially imagine extensive cooperation on the scale of hundreds to a few thousand people being facilitated by kinship—and cultural mechanisms for transubstantiating non-kin into kin—even at the same time as hunter-gatherers may be generally living with and marrying people who are not their genetic relatives (Hill et al. 2014; Ringbauer et al. 2021).

Acknowledgements

I would like to thank the workshop organizers for the opportunity to reflect on these ideas. Jamie Jones, Peter Collings, John Bunce, Andreas Womelsdorf, Willem Church and other conference participants provided thoughtful comments and discussion on earlier drafts of the paper. Opportunities to conduct fieldwork in Parnngurr and Ulukhaktok with Rebecca and Doug Bird and Peter Collings, and work with Nicholas Blurton Jones on Hadza census data, have provided an important comparative basis for this work. Most of all I would like to thank the families in Kangiqsujuaq who have let me "be with" them.

Funding

My research in Kangiqsujuaq has been funded by the Social Sciences and Humanities Research Council of Canada (752-2010-1089), the NSF Office of Polar Programs (PLR-13903874, P0106491), and the Max Planck Institute for Evolutionary Anthropology.

References

Balikci, A. 1964. *Development of Basic Socio-economic Units in two Eskimo Communities.* Anthropological Series No. 69. National Museums of Canada, Ottawa.

Bennett, J. and Rowley, S., (eds.). 2004. *Uqalurait: An Oral History of Nunavut.* McGill Queen's University Press.

Bird, D. W., Bird, R. B., Codding, B. F., and Zeanah, D. W. 2019. Variability in the organization and size of hunter-gatherer groups: Foragers do not live in small-scale societies. *Journal of Human Evolution*, 131: 96-108.

Bird-David, N. 2017. *Us, Relatives: Scaling and Plural Life in a Forager World.* University of California Press, Berkeley.

Bliege Bird, R., Ready, E., and Power, E. A. 2018. The social significance of subtle signals. *Nature Human Behaviour,* 2(7): 452.

Blurton Jones, N. 2016. *Demography and Evolutionary Ecology of Hadza Hunter-gatherers.* Cambridge University Press.

Bodenhorn, B. 2000. 'He used to be my relative: Exploring the bases of relatedness among Iñupiat of Northern Alaska. In *Cultures of Relatedness: New Approaches to the Study of Kinship* (ed.), J. Carsten, pp. 128-148. Cambridge University Press, Cambridge.

Boyd, R. and Richerson, P. J. 2020. *Large-scale cooperation in small-scale societies.* Paper presented at the 89th Annual Meeting of the American Association of Physical Anthropologists.

Briggs, J. 1970. *Never in Anger: Portrait of an Eskimo Family.* Harvard University Press, Cambridge, Massachusetts.

Burch, E. 1998. *Iñupiaq Eskimo Nations of Northwest Alaska.* University of Alaska Press, Fairbanks.

Collings, P. 2011. Economic strategies, community, and food networks in Ulukhaktok, Northwest Territories. *Arctic*, 64: 207-219.

Collings, P. 2014. *Becoming Inummarik: Men's Lives in an Inuit Community.* McGill-Queen's University Press, Montreal.

Collings, P., Pearce, T., and Kann, J. 2017. "We don't know anything about whales:" Ecological knowledge and ways of knowing in Ulukhaktok, Northwest Territories, Canada. *Arctic Science*, 4(3): 223-241.

Csonka, Y. 1993. Samuel Hearne and Indian–Inuit hostility. *Polar Record*, 29(169): 167-167.

Csonka, Y. 1999. A stereotype further dispelled: Inuit-Dene relations west of Hudson Bay, 1920-1956. *Études/Inuit/Studies*, 23: 117-144.

Damas, D. 1969. *Characteristics of Central Eskimo Band Structure.* Bulletin No. 228. National Museum of Canada, Ottawa.

Damas, D. 2002. *Arctic migrants/Arctic villagers: The Transformation of Inuit Settlement in the Central Arctic.* McGill-Queen's University Press.

Darwent, J. and Darwent, C. M. 2014. Scales of violence across the North American Arctic. In M.W. Allen and T.L. Jones (eds.), *Violence and Warfare among Hunter-Gatherers*, pp. 182-203. Routledge.

Dorais, L.-J. 2020. *Words of the Inuit: A Semantic Stroll Through a Northern Culture.* University of Manitoba Press.

Flora, J. 2019. *Wandering Spirits: Loneliness and Longing in Greenland.* University of Chicago Press, Chicago.

Fossett, R. 2001. *In Order to Live Untroubled: Inuit of the Central Arctic 1550 to 1940.* University of Manitoba Press, Winnipeg.

Graburn, N. H. H. 1969. *Eskimos without Igloos. Social and Economic Development in Sugluk.* Little, Brown and Company, Boston.

Hill, K. R., Wood, B. M., Baggio, J., Hurtado, A. M., and Boyd, R. T. 2014. Hunter-gatherer inter-band interaction rates: Implications for cumulative culture. *PloS ONE*, 9(7):e102806.

Jessen Williamson, K. 2011. *Inherit my heaven: Kalaallit gender relations.* INUSSUK–Arctic Research Journal.

Mauss, M. and Beuchat, H. 1904. Essai sur les variation saisonnières des sociétés eskimos: Étude de morphologie sociale. *L'Année Sociologique*, 9: 39-132.

Mitchell, M. 1996. *From Talking Chiefs to a Native Corporate Elite: The Emergence of Class Distinction among Inuit.* McGill-Queen's University Press, Montreal.

Morin, F. and d'Anglure, B. S. 1995. L'ethnicité, un outil politique pour les autochtones de l'Arctique et de l'Amazonie. *Études/Inuit/Studies*, 19: 37-68.

Nagy, M. 2006. Time, space, and memory. In P.R. Stern and L. Stevenson (eds.), *Critical Inuit Studies: An Anthology of Contemporary Arctic*, pp. 71-88. University of Nebraska Press, Lincoln and London.

Payne, F. F. 1889. Eskimo of Hudson's Strait. *Proceedings of the Canadian Institute, Third Series*, 6: 213-230.

Rasing, W. 2017. *Too Many People. Contact, Disorder and Change in an Inuit Society 1822-2015*. Nunavut Arctic College Media, Iqaluit, Canada.

Ready, E. 2018. Who, being loved, is poor?: Poverty, marriage, and changing family structures in the Canadian Arctic. *Human Organization*, 77(2): 122-134.

Ready, E. 2019. Why subsistence matters. *Hunter Gatherer Research*, 3(4): 635-649.

Ready, E. and Power, E. A. 2018. Why wage-earners hunt: Food sharing, social structure, and influence in an Arctic mixed economy. *Current Anthropology*, 59(1): 74-97.

Ringbauer, H., Novembre, J., and Steinrücken, M. 2021. Parental relatedness through time revealed by runs of homozygosity in ancient DNA. *Nature Communications*, 12(1): 1-11.

Saladin d'Anglure, B. 1967. *L'organisation sociale traditionnelle des esquimaux de Kangirsujuaaq*. Centre d'Études Nordiques, Québec.

Schneider, L. 1985. *Ulirnaisigutiit. An Inuktitut-English Dictionary of Northern Québec, Labrador, and Eastern Arctic dialects*. Les Presses de l'Université Laval, Québec.

Stupart, R. F. 1886. The Eskimo of Stupart Bay. *Proceedings of the Canadian Institute, Second Series*, 22(146): 95-114.

Trott, C. 2005. *Ilagiit and Tuqluraqtuq: Inuit understandings of kinship and social relations*. First Nations, First Thoughts conference, Edinburgh.

Wiessner, P. 2002. Hunting, healing, and hxaro exchange. a long-term perspetive on !Kung (Ju/'hoansi) large-game hunting. *Evolution and Human Behavior*, 23: 407-436.

Comment by Brian Codding

Focused on an ethnographic study with Inuit living along Hudson's Strait coast of Nunavik, referred to as Tarramiut, Ready offers a keen fine-grained perspective that highlights something absolutely central to human social organization, but often obscured in cross-cultural analysis, and invisible to archaeologists: all organization comes down to personal relationships. In doing so, Ready illustrates how the same individual practices which make relationships at the "tiny scale" of local organization, aggregate to form large-scale social organization, resulting in ephemeral constellations of individuals drawn from non-overlapping networks. This helps dispel three common misconceptions: first, that different mechanisms organize social units from one scale to the next; second, that social groups are cohesive units at any one scale; and third, that a society can be ascribed as "small" or "large" scale.

Regarding the first point, Ready discusses how kin concepts among Tarramiut are designed to emphasize that individuals are component parts of each other in family relations, yet these same concepts also function to extend social networks through practices like customary adoption, naming after others ("name-soul"), and staying with others; all of which allow one to extend relations beyond the local community to "make new family". Repeated throughout one's life, and maintained through "being together" in ways as mundane as informal visits to another's home, these practices result in expansive kin networks unique to each individual. Thus, the same practices that build social relations among individuals in a nuclear or extended family are applied to others in the same society, resulting in large-scale organization. These processes do not seem unique to Tarramiut life today, but appear to be part of a long-standing tradition.

On the second point, Ready illustrates that today, and in the past, groups which convene for one reason or another are "ephemeral constellations" of individuals drawn from a subset of each individual's larger network. This is a critical lesson, especially for cross-cultural and archaeological studies, which often mistakenly assume that observed or inferred groups are cohesive units. Instead, they should not be thought of as meaningful units of inquiry, but ephemeral expressions of relations among individuals who convene for a specific set of purposes in a specific place and time.

Taken together, these observations help convey why it might not be meaningful to ascribe labels such as "small-scale" or "large-scale" to human societies. If the mechanisms that build relations can both construct a nuclear household and build expansive interaction networks, and if any observed grouping is merely an ephemeral constellation of individuals connected through these mechanisms, then any society has the capacity to be "small" or "large", and may be both simultaneously.

In a volume on the scale of hunter-gatherer society that includes contributions across ethnographic, cross-cultural, and archaeological scales, Ready reminds us that all social relations come down to simple concepts that help people "be together". This should remind us that the same mechanisms can build small and large scales of social interaction, that individuals convened at any one time and place are not necessarily a cohesive unit, and that polarizing labels hide important patterning meaningful to the individuals who live in any one society.

Mikea, Malagasy, or hunter-gatherers?
Scale, ethnicity, and cultural groups in ethnographic description and ethnological analysis

Bram Tucker

Introduction

Are ethnic units also cultural and sociopolitical units? Barth (1969a) argued that they are not. However, some recent cultural evolutionary studies argue that ethnicity may function to facilitate within-group cooperation and between-group competition, referred to as parochial altruism (Choi and Bowles 2007; García and van den Bergh 2011; Handley and Mathew 2020; Jones 2018). Ethnographers have historically treated ethnicity and culture as equivalent with assertions that X people have particular beliefs, habits, customs, etc. In this paper I explore the ramifications of scale in ethnographic description and generalization, with a focus on my research participants in southwestern Madagascar, whom I usually label with the ethnonyms Mikea, Masikoro, and Vezo, or with the anthropological categories of hunter-gatherers, farmers, and fishermen. These are people who refer to themselves by these same ethnonyms or hyphenated combinations of terms (Masikoro-Mikea, Vezo-Mikea), or as Malagasy, a term refering to all peoples of Madagascar, or by village or clan affiliations. By contrasting evidence from my research (Tucker et al. 2021) with a study by Handley and Mathew (2020) about East African herders, I argue that the appropriate scale for ethnographic description may depend on patterns of similarity and difference in shared cultural traits and social networks, and these may be related to, or independent of, historically constituted ethnonyms. Careful thought is required to avoid scalar errors of over-particularization and exoticism (which I call Type 1 scalar errors) and over-generalization and stereotyping (Type 2 scalar errors). Because "ethnicity" is not just one "thing," ethnicity is not always the proper scale for ethnographic description.

I begin this exploration of scale in 2012, when the BBC News website posted a story about threats to the critically endangered spider tortoise (*Pyxis arachnoides*). The article argued that Madagascar's Mikea hunter-gatherers pose a significant threat to the tortoise by over-hunting (Barley 2012). The story made some significant errors. It erroneously referred to Mikea people as "a nomadic tribe," and it repeated the tired, ethnocentric narrative that ignorance and poverty drive traditional people to overexploit endangered natural resources (cf. Kull 2000; Scales 2012 who challenge this narrative). But a central claim of the article is at least partially factual: Mikea people do catch tortoises and bake them in hot coals to eat the meat inside. Or so I have been told. During 25 years of fieldwork with Mikea I have never witnessed the practice and it is unclear to me whether it occurs with sufficient frequency to constitute a threat.

I start with this BBC article about tortoise hunting because it makes a critical and potentially dangerous error of scale, a type of error that is common in media and social science descriptions of peoples in the rural Global South. Whereas the BBC story was correct that Mikea hunt and eat tortoises, the claim is misleading because "Mikea" is the wrong scale at which to ascribe the practice. Many, or maybe most, Malagasy peoples hunt, cook, and eat tortoises in the same way, whether they self-identify as Masikoro, Vezo, Bara, Mahafale, Tanosy, Tandroy, etc. Tortoise eating should more properly be ascribed to some or all Malagasy, where "Malagasy" or *olo Gasy* is a salient national identity term encompassing all of Madagascar's 25 million people. The BBC article's claim is potentially dangerous because it places the blame for overhunting on a subset of the likely "culprits," who happen to be among the poorest and least able to defend themselves in public narrative or legal tribunal. The accusation that Mikea are to blame for overhunting endangered species could invite conservation-minded project planners and policymakers to unfairly limit Mikea people's access to the wild foods they need while not placing similar limits on their wealthier tortoise-eating neighbors.

I call this a "Type 1" scalar error, in playful reference to type 1 errors in frequentist statistics. A Type 1 scalar error occurs when writers ascribe traits to a small social unit that are, in reality, shared by the larger population, of which the smaller unit is but a subset. The small social unit is often labeled with an ethnic term. The harm of Type 1 errors is that they make minority groups stand out as exotic, while making group boundaries seem more concrete than they may be on the ground.

Mikea may be particularly vulnerable to harm from Type 1 scalar errors. Because Mikea are rumored to be Madagascar's only hunting and gathering population, they are consistently presented as primitive people, clothed in familiar idioms of both noble savagery and mysticism, as well as sub-humanness and backwardness. I commonly hear from urban Malagasy that Mikea are African pygmies without language, or that they are invisible, or that they are the last remaining survivors of Madagascar's original people who occupied the island before the arrival of proto-Malagasy from Indonesia (Poyer and Kelly [2000] report hearing similar stories). Early ethnographers claimed that Mikea live in a "repulsive" desert environment (Dina and Hoerner 1976: 275) of "thorns" without consuming water (Molet 1958, 1966). Popular journalistic accounts describe Mikea as nomads in harmony with nature who live in rudimentary huts without use of money or markets; and as people threatened by rapacious Malagasy farmers who ravage their forests for agriculture (Mouyon and Francelle 1999; Rarojo 1999). The World Bank classified Mikea as Madagascar's only indigenous people (Huff 2012). Documents instrumental in the creation of the Mikea Forest National Park stated that there are fewer than 1000 Mikea people living in a handful of villages, and that Mikea life is intimately tied to the cult of the ancestors and animistic rites (Repoblikan'i Madagasikara 2010: 20-21).

Some of these statements are absurd: Mikea are not pygmies; like all humans, they talk, drink water, participate in new economic opportunities, and have positive and negative effects on the environment. Genetic evidence demonstrates that Mikea share historical origins with other Malagasy (Pierron et al. 2014; Razafindrazaka et al. 2010), which is consistent with Mikea oral histories that tell of their shared ancestry with neighboring Masikoro and Vezo people (Tucker 2003). There are many more than 1000 Mikea (probably more than 10,000) but the number depends on some fuzzy definitions.[1] A few other statements result from Type 1 scalar errors: Mikea are not the only

1 The 1000 person Mikea estimate seems to be a sum of people in the Namonte Basin, Bedo, and a few other large forest communities. But on the edges of the Mikea forest there are a series of villages where many or most people call themselves Mikea (or Masikoro-Mikea, or Vezo-Mikea), many of which were founded in the early twentieth century as a result of French colonial relocation and villagization projects. Some of these villages are large; Magnono, Andohasakoa, Vorehe, and Bevondro each have several thousand Mikea inhabitants, while hundreds more Mikea live in villages such

Malagasy to live in "rudimentary huts;" nearly identical reed-thatched structures house wage workers throughout urban Toliara. Most rural Malagasy hunt and gather, in addition to farming and herding. In Toliara I routinely drink beer with two old friends, a Tesaka retired hotel guard from eastern Madagascar and a Tandroy rickshaw cyclist from southern Madagascar, who, after a Three Horses Beer and a pack of cigarettes, inevitably wax about their childhood adventures chasing *tandrake* (*Tenrec ecaudatus*) and digging *ovy* tubers (*Dioscorea acuminata*), the same wild prey that Mikea pursue. Mikea are hardly unique in their devotion to 'the cult of ancestors' and 'animistic rites'; most Malagasy, whether Merina, Betsileo, Tagnala, or Tankarana venerate ancestors and forest spirits (Mack 1986; Middleton 1999). However, when ascribed to Mikea, such behaviors make Mikea seem different and exotic, whereas much of this description fits most Malagasy living in towns and cities across the island nation.

Hunter-gatherers are particularly vulnerable to a second, "Type 2" scalar error, which occurs when writers generalize observations from a small population to a larger category of which they are a supposed subset. This error is commonplace in news media descriptions. A casual internet search for "hunter-gatherer" news turns up a series of remarkable claims: "what a hunter-gatherer diet does to your body in just three days" (Spector 2017), "hunter-gatherers agree on what is moral, but not on who is moral" (Science Daily 2021), and "hunter-gatherers sit as much as us, but how they sit makes all the difference" (Dockrill 2020). There are an estimated 5,000,000 hunter-gatherers in the world today, living from the arctic to the tropics, with diverse diets, concepts of morality, and sitting postures (Lee and Daly 1999; Kelly 2013). But all three of these news stories generalize about all "hunter-gatherers," including those in the distant past, with observations from a single contemporary population, Hadza of Tanzania, and from the limited subset of Hadza recruited for each research project. The original research these news articles refer to make more precise scalar claims: neither K. Smith and Apicella (2020)'s study of Hadza morality, nor Raichlin et al.'s (2020) study of Hadza sitting, generalize their findings to all foragers. Some scholarly work generalizes about hunter-gatherers from just one or two populations. For example, Majid and Kruspe (2018) conclude that "hunter-gatherer olfaction is special," based on data from Malaysian Semaq Beri foragers in contrast to

as Ihotre, Antsakoamarovitike, Afeza, Befandefa, Ankindranoke, Agnolignoly, etc. This partial list excludes the southern half of the Mikea Forest.

their horticulturalist neighbors; and D. Smith et al. (2014) speculate about the significance of "hunter-gatherer story telling" for cooperation, based only on stories from Agta of the Philippines. But many contemporary studies that make general claims about "hunter-gatherers" involve some degree of cross-cultural comparison, seemingly to avoid Type-2 errors (e.g., Bird et al. 2019; Bird-David 2017; Hamilton et al. 2007; Hill et al. 2011; E. A. Smith et al. 2010).

Hunter-gatherers are vulnerable to Type 2 scalar errors because of lingering nineteenth century notions that foragers are relics unchanged since humanity's earliest stage of cultural evolution in the Pleistocene. Although anthropologists have opposed this social evolutionist worldview for the past half century (Barnard 1999; Schrire 1984), it still occupies popular conceptions of human history, and occasionally slips into academic work. It is hard to imagine headline statements that *farmers'* or *wage workers'* diets, moral concepts, or ways to sitting have the same relevance to human nature and our evolutionary past.

In this chapter I consider whether ethnographers commit Type 1 or Type 2 scalar errors when we generalize findings to the level of the ethnic group. Generalizing observations to ethnicities is an old practice in anthropology, as demonstrated by the volumes that populate anthropologists' bookshelves, with titles like *The Yanomamo* (Chagnon 2012), *The Canela* (Crocker and Crocker 2004), *The Tiwi of North Australia* (Hart and Pilling 1963), *The Bolivian Aymara* (Buechler and Buechler 1970), etc.; and, as demonstrated in cross-cultural studies, in which the datapoints are "societies" with unique social and cultural traits that are labeled with ethnonyms (e.g. Borgerhoff Mulder 2009; Ember 1978; Henrich et al. 2005; Murdock 1967). More casually and commonly, ethnographers routinely state that we work with this-or-that ethnic group, or that the ethnic group we study has this-or-that set of customs or beliefs. Generalizing ethnographic observations to ethnicities implies that the world's peoples fall into natural, discrete, comparable cultural units that correspond to ethnic boundaries. Yet we know that ethnic identities are often flexible and negotiated (Astuti 1995; Linnekin and Poyer 1990); that ethnonyms are often imposed by outsiders during processes of conquest and colonization (Mafeje 1971; Iliffe 1979; Southall 1970); that ethnic groups vary in scale, from nano- to nation (Bird et al. 2019; Bird-David 2017); and that ethnicity is only one among a host of identities that are imposed upon and adopted by the subjects of research, alongside gender, nationality, occupation, and residence, and kin group (Barth 1969a, b). Generalizing to the ethnic level constitutes a Type 1 error if the social or cultural traits we describe are common among a larger populace, perhaps defined by regional geography, language, nationality. Generalizing to the ethnic level constitutes a Type 2 error if the traits we describe are actually particular to individuals, families, clans, neighborhoods, villages, areas, or genders. These social scales, from family to village to ethnicity to nation, may have a hierarchical, nested structure, or they may be cross-cutting and negotiable.

I begin by re-visiting conflicting claims in the literature about the relationship between ethnicity and culture. Fifty years ago, Barth (1969a) argued that ethnic units should not be considered cultural or social units. For simplicity, I abbreviate this argument as *ethnicity ≠ culture ≠ society*. Barth and his intellectual descendants argue against the received wisdom that the world's people constitute an array of discrete cultural units who bequeathed their heritage faithfully and linearly across generations since the beginning of time. Rather, ethnicity and identity are social facts (*sensu* Durkheim 1982[1895]) that we collectively imagine into being, and that we are constantly reimagining and renegotiating, that may correspond poorly with actual patterns of cultural agreement and social organization.

Then I discuss recent work by cultural evolutionary scholars who argue that ethnic units may be cultural and social units; that "ethnicities" could have evolved through cultural group selection to divide humans into internally-cooperative and externally-competitive groups, a pattern called parochial altruism (Choi and Bowles 2007; García and van den Bergh 2011; Handley and Mathew 2020; Jones 2018). I abbreviate this argument as *ethnicity = culture = society*. This perspective views ethnicities like sports teams, who mark inclusion with team colors, within which teammates, bound by their cultural similarities, work together to advance their survival and to defeat other, culturally-foreign teams.

I contrast two published studies, one supporting *ethnicity = culture = society* with evidence from east African herders (Handley and Mathew 2020), and the other, coauthored by myself and colleagues, supporting *ethnicity ≠ culture ≠ society* in southwestern Madagascar (Tucker et al. 2021). I discuss the historical and geographical reasons why ethnicity is a different kind of thing in these two places. Then I conclude with some thoughts about best practices when generalizing across scales. I suggest that labeling samples with ethnonyms may be unwise even when evidence supports *ethnicity = culture = society*. We should be particularly cautious in making scalar claims when there is risk that Type 1 or Type 2 errors could cause harm, such as when describing behaviors our audience might associate with primitiveness. Generalizations about hunter-gatherers ideally require data from a large and preferably representative sample of foraging populations as well as a non-foraging control groups.

Barth: Ethnicity ≠ Culture ≠ Society

Studies of ethnicity and identity routinely cite Barth's (1969b) edited volume as the starting point for modern research on the topic. In the introduction, Barth (1969a) presents the old-school "ideal type" description of ethnicity:

> "Practically all anthropological reasoning rests on the premise that cultural variation is discontinuous: that there are aggregates of people who essentially share a common culture, and interconnected differences that distinguish each such discrete culture from all others. Since culture is nothing but a way to describe human behaviour, it would follow that there are discrete groups of people, i.e., ethnic units, to correspond to each culture." (Barth 1969a: 9)

Two pages later Barth (1969a: 11) continues, it is "not so far removed in content from the traditional proposition that a race = a culture = a language and that a society = a unit which rejects or discriminates against others." He then proceeds to dismantle this old-school ideal type, arguing that ethnic boundaries often facilitate social ties that cross boundaries. Cross-boundary social relations may be just as important as coethnic relations, and are not necessarily agonistic (think of trade, for example). He then argues that ethnicities are not culture-bearing units. People of the same ethnicity occupying different ecologies are likely to have different cultural traits. Pathan of Afghanistan and Pakistan perceive unity among fellow Pathans, even though the cultural traits of northern and southern Pathan are quite different, and regionally, Pathan may be more culturally similar to neighboring non-Pathan than to distant coethnics. Pathan perceive Pathan unity around a small assortment of seemingly arbitrary cultural traits.

Cultural evolutionary arguments for *Ethnicity = Culture = Society*

Now let us fast-forward to the first decades of the 21st century, and recent arguments about ethnicity and parochial altruism by scholars of cultural evolution. Parenthetically, contemporary cultural evolutionism is completely unrelated to the racist, colonial, Victorian cultural evolutionism of Herbert

Spencer, Edward Tylor, and Lewis Henry Morgan. It also has very little to do with genes. Modern evolutionary thought starts with Darwin's tenants that there is variation within and among populations, that some of this variation is heritable, and that some variants are more likely to survive to reproduce within particular environments. Whereas biologists commonly apply Darwin's tenants to genetic inheritance and biological diversity, Darwin's tenants apply equally well to cultural inheritance and behavioral diversity. Culture varies. Culture is heritable, through active and passive forms of teaching and learning. Culture delivers survival and reproductive outcomes. Unlike genes, we acquire culture continuously throughout our lives from many sources. Cultural information rarely consists of discrete units, but consists instead of knowledge sets, associated, for example, with subsistence, religion, or identity. These knowledge sets contain norms for rewarding compliance and punishing deviance, which operate to make the knowledge seem normal, moral, or inevitable (for an accessible introduction to this theory, see Richerson and Boyd 2005).

A major research question for contemporary studies of cultural evolution is how to explain how people get along with one another in large scale societies constituted by anonymous strangers. When two people meet, they must first solve a series of coordination problems, such as how to greet one another. A handshake, hug, or kiss on the cheek work equally well, so long as both people share the same expectation, and there is no benefit to transgressing the standard. Strangers may also have to solve cooperation problems involving sharing or helping, that are costly to perform, beneficial to receive, and prone to cheating (non-reciprocation). As the argument goes, in small-scale, kin-based societies, coordination and cooperation problems are easily solved because people interact frequently and remember each other's past behavior. But in large-scale societies, one cannot be sure of a stranger's history of past transgressions, or even what they consider to be transgressive behavior.

Ethnicity can solve problems of coordination and cooperation at large scales if visible, ethnic practices communicate invisible commitments to social norms (McElreath et al. 2003). If ethnic markers, visible tags such as clothing, hairstyle, or dialect, coevolve with social norms, then one may know from glancing at a stranger's clothes and hair what rules they follow, and interact with them accordingly (McElreath et al. 2003; Riolo et al. 2001). If ethnic markers co-evolve with cultural information and social norms, then ethnic groups, bounded as they are by markers, are also likely to be cultural and social groups. Coethnics share the same coordination norms for things like greetings. Members of an ethnicity are incentivized to avoid cheating in cooperation problems because of fear of punishment, ostracization, and the loss of group membership and shares of collective gains. The result is within-group social cohesion (Moya and Boyd 2015). Ethnic beliefs and practices may strengthen outgroup antagonism, because warfare is a cooperative act in which coethnics reward each other for victory and punish cowards and defectors (Mathew and Boyd 2011). The result of this "parochial altruism" may be that ethnic groups rather than individuals compete for survival, a type of cultural group selection (Choi and Bowles 2007; García and van den Bergh 2011; Handley and Mathew 2020; Jones 2018).

Example of *Ethnicity* = *Culture* = *Society*: East African herders

Handley and Mathew (2020) offer a formal definition of "cultural unit" as the social scale at which two individuals from different groups are most likely to disagree about cultural norms. Differentiated cultural units are analogous to differentiated genetic populations and can be quantitatively evaluated using the same math, Wright's fixation index (F_{ST})

Handley and Mathew (2020) test key predictions of parochial altruism using data from four East African pastoralist populations, Samburu, Borana, Rendille, and Turkana. To test whether cultural norms differ more between ethnic groups than between territorial or clan subsections of an ethnic group, they asked 793 individuals whether they agreed or disagreed with a series of 49 normative statements (e.g., "A woman can only joke with a man from her husband's ageset"). They found the greatest variation in norms (the greatest cultural F_{ST} or CF_{ST}) was between pairs of ethnic groups: Samburu, Borana, Rendille, and Turkana (mean CF_{ST} = 0.152). There was much less variation in norms between subsections of ethnic groups (mean CF_{ST} among Turkana territorial sections = 0.030; among Borana clans = 0.003). This evidence suggests that Samburu, Borana, Rendille, and Turkana ethnic units correspond to social and cultural units.

To test whether cooperation is more likely among sets of people with more similar social norms, Handley and Mathew (2020) asked the same sample to respond to a series of hypothetical vignettes involving helping or not helping others from the same or different groups. They found a greater willingness to cooperate with those from groups with more similar norms. Thus, ethnic units appear to correspond to socio-political groups.

These findings are remarkably consistent with colonial era anthropologists' descriptions of hierarchical, nested groups of ethnicities and ethnic subunits. For example, in Evans-Pritchard's (1940) description of Nuer, he classes Nuer and Dinka, and Shilluk and Luo, as two branches of a larger category of Nilotes. He subsequently subdivides Nuer into nested categories at a cascade of levels from tribes to primary, secondary, and tertiary tribal sections to villages, each with its own territory, character, and identity. This is the hierarchical nested scalar model of cultural and identity which Barth (1969a) criticizes.

Example of *Ethnicity ≠ Culture ≠ Society*: Mikea, Masikoro, and Vezo of southwestern Madagascar

My colleagues from the University of Toliara[2] and I (hereafter, "we") performed similar data collection and analyses as Handley and Mathew and arrived at different conclusions (Tucker et al. 2021). For our study, we wanted to know how well Mikea, Masikoro, and Vezo could classify one another into ethnic categories based on visual cues, and whether they prefer to cooperate with coethnics.

Identity in this part of Madagascar is complicated and has been the subject of much research (Astuti 1995; Astuti et al. 2004; Poyer and Kelly 2000; Tucker et al. 2003; Yount et al. 2001). Ask just about anyone in the region what Mikea, Masikoro, and Vezo are, and you will probably get an answer such as these statements, made by women in the context of focus group discussions about ethnicity in 2006:

> "Mikea live in the forest, and they sell what they gather from the forest. Masikoro live in the interior. They cultivate rice, manioc, sweet potato. Vezo do their livelihoods at sea."

> "Mikea live in the forest; they know how to collect honey and tenrecs, and hunt wild bushpig. Masikoro are people who practice the circumcision ceremony [for boys]. Vezo do not circumcise; they do their livelihoods in the sea."

---

2 Special thanks to Dr. Tsiazonera, Dr. Jaovola Tombo, Patricia Hajasoa, Soanahary Gérard, Rolland Lahiniriko, Angelah Halatiana Garçon, Gervais Tantely, Théodore Tsitindry Ramanovontsoa, Jean-Claude Alhayess, Repapa Pamphil de la Patience, and Eric Rambeloson.

These statements suggest that Mikea, Masikoro, and Vezo are mostly livelihood distinctions, so that Mikea are hunter-gatherers, Masikoro are farmers, and Vezo are fishers. Astuti (1995) explored the ramifications of this "identity by doing" among Vezo and their neighbors in the Menabe Region, 150 km north of our field sites. She documents that children born of Vezo parents are not considered to be Vezo until they learn to "struggle with the sea," and that adult Masikoro can become Vezo by moving to the coast and learning to fish or sail.

Although we have heard similar narratives in our field sites (Poyer and Kelly 2000; Yount et al. 2001), we have also noticed frequent mismatches between identity and livelihood, including whole villages where people self-identify as Mikea despite farming or fishing for a living, villages of farming Vezo, and villages of Masikoro who fish. Most Vezo villages contain immigrants who practice Vezo lifeways but are nevertheless called Masikoro, and Vezo may farm the savanna and not be considered Masikoro by their neighbors. When we ask people to explain these apparent mismatches between ethnicity and occupation, we hear a second narrative, that identity is inherited lineally from ancestors. Mikea are those who venerate ancestors who resisted the Andrevola kings that ruled the region before French colonization by hiding in the forest; foraging is a symbol of resistance and independence and not necessarily a specialization. Masikoro venerate ancestors who were vassal to the kings; crops and cattle symbolize wealth and strength. Vezo remember ancestors who resisted royal dominion by sailing away to sea (Tucker 2003). Children acquire these identities during rites of filiation (*soroanake*), when their formal relationship with ancestors begins.

Given the competing narratives of what makes someone Mikea, Masikoro, and Vezo, identity fluidity, and routine, peaceful inter-ethnic interaction for trade, marriage, and ritual, we wondered whether people could actually discriminate one another by ethnicity just by their appearance (Tucker et al. 2021). We took photos of 132 Mikea, Masikoro, and Vezo adults (we call the photographed subjects "alters") standing alone against a blank background. Then we showed these photos to 355 Mikea, Masikoro, and Vezo living 100 km or more away (the "judges").

In the first experiment, judges were asked to classify alters who were photographed in their everyday clothes, without objects in their hands, in an upright pose. The judges successfully identified the alters 65% of the time, which is much greater than the background guessing rate (33%). This indicates that Mikea, Masikoro, and Vezo do send and receive signals marking their ethnicity. In the second experiment we asked judges to classify photos of alters who were specifically posing as a member of their ethnic group. Mikea alters donned tattered clothes and held the digging tools and net bags they use in foraging; Masikoro women tied sarongs (*lamba hoany*) high on their chests and across their breasts while men wore capes (*lamba be*) and hats, holding spades and plows; and Vezo women tied their sarongs low, and Vezo women and men held fishing lines and lures, fishing spears, masks, fins, and in one case, a dead squid. Although judges were more successful classifying these photos (77% success), the difference is not statistically significant, suggesting that southwestern Malagasy may be limited in their ability to purposefully improve the clarity of their ethnic marks. In the third experiment, judges classified photos of alters recruited from busy marketplaces, venues where Mikea, Masikoro, and Vezo interact. Interestingly, judge's success rate matched that of the probability of a guess. These experiments indicate that Mikea, Masikoro, and Vezo do perceive borders separating these identities, although they seem to drop the borders when meeting to trade.

The fourth experiment was a bit different. Judges were told a hypothetical vignette about a wage labor opportunity to the north, where the boss hired people by teams. Whichever team cooperated the best got double salary, and whichever team cooperated the worse got half salary. We then asked, who, among the photographed alters, would you most want on your team, and who do you least trust to cooperate? If ethnic boundaries are social boundaries marking discrete transitions between sets of social norms, then we would predict that the judges would prefer to cooperate with coethnics. This was not the case; judges were equally likely to classify coethnics as cooperative and as untrustworthy. Judges explained their choices with reference to the alters' appearance, work ethic, and personality. Only one out of 90 judges mentioned ethnicity as a reason to mistrust a coethnic, and three out of 90 cited a preference to work with someone of a different ethnicity. Ethnic boundaries do not appear to be social boundaries.

In a separate series of semi-structured interviews conducted with 30 Mikea, Masikoro, and Vezo in three villages, we asked whether people of their ethnicity cooperate best with coethnics or with people from neighboring ethnicities. Roughly half (16 out of 30) said they worked best with coethnics, citing similar knowledge, ideas, livelihood strategies, personality, and sense of humor, and several said members of their ethnicity work best in solitude (N=2). A substantial minority expressed a preference for working across ethnic lines. The benefits of coethnic cooperation include a more diversified skillset, lower political tension (because coethnics tend to argue and compete about similar things), and cross-ethnic kindred in perspectives.

To discover whether ethnic boundaries correspond to cultural boundaries, we conducted two versions of a norms questionnaire with 150 people per version in two Mikea, two Masikoro, and two Vezo villages. We asked questions about social organization and gender (e.g., "is it normal to marry someone in your clan?"), ancestor veneration (e.g., "Is it normal to offer the ancestors a goat, wild tubers, or beans if you don't have access to a sacrificial cow?"), and taboos (e.g., "are you taboo for sheep?"). We purposefully chose items that we suspected might be different among ethnicities (e.g., we had heard that Vezo are less concerned with clan endogamy, Mikea routinely offer goats or beans in sacrifice to ancestors, and Vezo are universally taboo for sheep). Like Handley and Mathew, we calculated cultural F_{ST}, with the help of coauthor Erik Rinen. The results indicated that that there was similar variation in norms between villages as between ethnicities (Average CF_{ST} among pairs of ethnicities = 0.04 for social organization norms, 0.06 for ancestor norms, and 0.07 for taboos; average CF_{ST} among pairs of villages = 0.05 for social organization norms, 0.06 for ancestor norms, and 0.05 for taboos). Although we might have found greater ethnic differences with different questions or a larger sample of villages, these analyses suggest that ethnic boundaries do not enclose cultural differences.

East-African herders are different from southwestern Malagasy because ethnicity is not just one thing

Readers could argue that by comparing the findings of Handley and Mathew (2020) to my own study (Tucker et al. 2021) I am setting up a false comparison, because Turkana, Samburu, Borana, Rendille, are clearly not comparable ethnological units as Mikea, Masikoro, and Vezo. Turkana, Samburu, Borana, and Rendille and speak different languages, from two different language families; they migrated to their current territories from different directions; and they sometimes raid each other for livestock. By contrast, Mikea, Masikoro, and Vezo speak the same language, share historical origins, in many cases belong to the same clans, and depend on one another for trade, marriage, and ritual. But these differences are exactly my point. By ascribing ethnographic descriptions to ethnicities, anthropologists, journalists, and politicians speak as though these *are* comparable units. In this section I discuss some of the geographical and historical reasons why ethnicity is a different phenomenon in East Africa contrasted to Madagascar.

First, let us consider geography and deep time. The borders of Kenya are modern and arbitrary, whereas the borders of the island nation of Madagascar are unambiguous coastline. Kenya is a cradle of human evolution while Madagascar is among the last large landmasses to be occupied by humans (even with new evidence by Dewar et al. 2013). The result is that the modern borders of Kenya arbitrarily group together various peoples with diverse languages and cultures, who may find many reasons to see each other as different. By contrast, there is only one native language in Madagascar, Malagasy, spoken in different dialects by all 25 million inhabitants. Genetic studies find that despite the dual origins of the Malagasy population in Africa and Island Southeast Asia, there is low haplotype diversity and relatively even admixture of African and Asian genetic markers (Pierron et al. 2014; Razafindrazaka et al. 2010). Malagasy share many significant cultural practices and beliefs centered on ancestor veneration, cattle sacrifice, tombs, spirit possession, divination, and astrology (Mack 1986; Middleton 1999). Malagasy may be less likely than East Africans to see ethnic difference because they are less differentiated. Indeed, people in Madagascar habitually refer to Malagasy customs (*fomba gasy*), Malagasy knowledge (*fahaiza gasy*), and Malagasy food (*sakafo gasy*) rather than the customs, knowledges, and foods of smaller ethnic subunits, even when describing local practices that are not actually shared across the island.

Next, consider the ways that European colonial powers exploited social differences for political purposes. The Germans and British in East Africa and the French in Madagascar employed similar strategies of codifying racial and tribal boundaries and transforming these into colonial administrative units via policies of indirect rule, but with some different outcomes. European explorers assumed *à priori* that Africa's peoples fell naturally into racial and tribal categories, and then sought to document those categories whether they existed or not (Iliffe 1979; Mafeje 1971; Ranger 1993). Nineteenth century linguists observed the geographic distribution of grammars and vocabularies and, from them, invented stories of sequential invasions by races of increasing superiority: Bantu replacing San, Nilotes displacing Bantu, Arabs subjugating Nilotes (Gourevitch 1996). The list of supposed tribes generated by explorers included an odd collection of dissimilar categories: geographical names, kin groups, kingdoms, enemy's epithets, and catchall categories (Southall 1970, 1971). European powers transformed these newly discovered (invented) tribes into administrative units, and transformed whatever influential people they found, whether kings or clan heads or healers, into "chiefs," whom they incorporated into the colonial government hierarchy, where they were responsible for enforcing colonial policies of taxation, labor recruitment, and villagization (Hodgson 1999; McCabe 2004; Simpson and Waweru 2021). Through this process, Europeans may have brought into existence the hierarchical, nested sociocultural groupings that Evans-Pritchard (1940) and other colonial era ethnographers assumed were primordial.

In Madagascar, efforts to reify ethnic or tribal identities may not have been entirely successful. French colonial agents worked with a list of 18 supposed tribes generated by French explorer Alfred Grandidier and others (Kent 1970; Southall 1971). Grandidier's tribal map labels Madagascar's southwestern people as Sakalava, but the Mikea, Masikoro, and Vezo people that I work with do not seem to have ever used this term for themselves. This may be because the term Sakalava is a place name that refers to the region north of where my fieldwork occurs, or because southwesterners were never administered collectively as Sakalava. In the south, southwest, and west, indirect rule involved empowering the sons and grandsons of the last kings rather than ethnic representatives per se. Eggert (1986) met many people in the area labeled Mahafale who still had not heard that they were supposedly Mahafale, suggesting that some Malagasy did not know their supposed tribal affiliations until instructed by outsiders.

Published histories of the origins of east African herders seem consistent with parochial altruism. During the centuries before colonization, one branch of Nilotic speakers diverged to form Turkana, Karimojong, and Jie (Lamphear 1988; McCabe 2004), while another split to form Samburu and Maasai (Simpson and Waweru 2021), whereas Borana, Rendille, and Ariaal diverged after the rise of the Oromo kingdom in the horn of Africa (Schlee 1990), during a time of war over pasture and raiding for cattle.

These histories may be largely factual, but the question remains whether eighteenth century east Africans called themselves by these ethnic terms imbued with their contemporary meanings, and formed alliances and enmity along ethnic lines; or whether the ethnic terms were applied during after-the-fact twentieth century retelling because ethnic divisions had become meaningful in the colonial era. While there was inter-ethnic conflict in the precolonial era, there is also evidence of inter-ethnic cooperation. By the 18th century, Samburu, Rendille, and Borana formed "heterogenous, multilingual confederations" (Lamphear 1988: 31 cited in McCabe 2004: 49). Following a period of interethnic conflict associated with the growth of the Oromo polity, Samburu enjoyed a "pax Borana" with their neighbors (Schlee 1990). Some scholars have suggested that the warfare and raiding by herders witnessed by colonial officers could have been a recent reaction to the 1890s rinderpest epidemic that devasted herds, or a reaction to colonial intrusions and policies (McCabe 2004; Oba 2011). As a result, it is unclear to what degree the parochial altruism identified by Handley and Mathew is the cause of Turkana, Samburu, Borana, and Rendille ethnogenesis or the result of colonial policies of division.

Conclusions: At what scale should we generalize ethnographic descriptions to avoid Type 1 and Type 2 errors?

Although the scale of ethnographic representations was not the focus of Handley and Mathew's study of East African pastoralists nor my co-authored study in southwestern Madagascar, one could conclude from our studies that the proper scale of generalization should be the scale where there is the greatest between-group difference. In Handley and Mathew's East African Pastoralist example this seems to be the ethnic level, whereas in southwestern Madagascar it would be something larger. I endorse a qualitative application of this strategy, with some significant caveats.

One caveat is that even in cases where cultural knowledge and social structure do demonstrably cluster at ethnic levels, labeling cultures and societies with ethnonyms may still be unwise. This is because the practice encourages a casual "ethnicism" with the same dangers as everyday racism, as an anonymous reviewer of an earlier draft of this chapter suggested. That people who self-identify as Turkana agree more amongst themselves than with non-Turkana about men and women's joking relationships does not indicate that these norms are inherent to, or caused by, being Turkana, for regional agreement in norms could be coincidental to the ethnonyms and identities employed in the region. Nor do we know how change or loss of these norms might influence Turkana identity, if at all. Use of ethnic labels may encourage the general public to think that ethnic groups as primordial or essentialized populations. Exotic-sounding ethnonyms may conjure inaccurate stereotypes of primitivism.

A simple linguistic solution to ethnic labels would be to change statements such as "I study Mikea," to "I study people in southwestern Madagascar who self-identify as Mikea," followed by a description of what self-identifying as Mikea means. This relatively simple rephrasing indicates although my research subjects are Mikea, my findings do not necessarily apply to all Mikea.

This rephrasing also leaves open the possibility that my research participants may call themselves by other terms (which they do). This practice is standard among many anthropologists, but it is hardly universal across the social sciences, much less in public media.

Some writers continue to put the definite article before ethnonyms, so that they write about "the Mikea," "the Yanomamo," in the same way that politicians half a century ago talked about "the Blacks," "the gays," etc. Putting "the" before an ethnonym makes ethnicity a noun and an immutable category of matter. Dropping "the" leaves the ethnonym to function like an adjective, so that "Mikea" or "Yanomamo" is a property of a person, and just one property at that.

Cross cultural studies should refer to their samples with geographic locations rather than ethnonyms, although it may be appropriate to explain that the people from *X* sample self-identify as *Y*. This strategy avoids conflating the sample with an ethnicity, society, or culture, and it avoids presenting ethnic groups as comparable units of analysis.

It is probably impractical use quantitative measures of CF_{ST} as a guide to the scales for ethnographic generalization. For one thing, different cultural traits and social structures within the same populations may generalize at different scales. For example, some specific beliefs about forest spirits may be unique to some Mikea individuals or communities, whereas other beliefs endorsed by Mikea, such as the general belief that forest spirits exist and mediate between living supplicants and God the Creator, are common to all Malagasy, and perhaps beyond Madagascar as well. To calculate CF_{ST} for multiple social and cultural features across scales would require an overwhelming amount of data, from a plurality of people who call themselves Mikea, Malagasy, and perhaps from across the Indian Ocean Rim.

A qualitative application of the basic logic of CF_{ST} calculations, which states that we should generalize at the scale where beliefs and practices are shared, requires writers to have a general knowledge of regional cultural patterns. So rather than write, "Mikea believe that people with bad intent can harm others through the manipulation of magical objects," a more cautious statement would be that "many people in the study region, like their neighbors across much of rural and urban Madagascar and Africa, believe that people with bad intent can harm others through the manipulation of magical objects."

Obviously the second version of this statement is more complicated, and writing space is often limited. Thus, we should prioritize using cautious scalar statements when describing topics with the greatest chances of causing the dangers of Type 1 and Type 2 errors, specifically, exoticization and stereotyping. This includes cultural traits associated with primitivism such as sorcery, witchcraft, spirit possession, scarification, human sacrifice, skull deformation, genital and other body modifications, cannibalism, marriage-by-capture, and child marriage, but also, hunting and gathering, nomadism, chiefs, etc.

The appropriate scale of ethnographic generalization may not have a handy and convenient name, and the temptation to create new scalar names could lead to a counter-productive return to cultural area studies. I have provided evidence that Mikea, Masikoro, and Vezo share many cultural beliefs, but people in the region employ no umbrella terms to refer to these people. Some early writers referred to these people as "Fiheregnars" (Drury 1826[1729]) after the name for the Fiheregna region, but "Fiheregnars" lacks local salience. I use the admittedly awkward label "southwestern Malagasy." Ralph Linton (1928) suggested that Mikea, Masikoro, and Vezo share many cultural traits with other western and southwestern Malagasy, which he labels "cultural area III," which corresponds to the arid parts of Madagascar. This culture area approach conflates environment with culture and erases as much variation as it labels.

Sometimes our research goals require generalizations at scales such as ethnicities, nationalities, or anthropological categories such as "hunter-gatherers." In these cases, it is important to draw conclusions from a representative sample of both members and non-members of the category we are generalizing about. For example, I have argued elsewhere that anthropologists may be unduly fixated on food sharing as a trait common to hunter-gatherers, when actual sharing attitudes and behaviors vary considerably among foraging populations and farmers and herders also share food with similar apparent generosity as some foragers (Tucker 2019). The question remains how many samples of foragers (and of a non-forager control group) are required to generalize about foragers, seeing as a sample representative of foragers and non-foragers across time and space would be challenging to acquire. The two articles that inspired the Scale Matters workshop generalize to all foragers from one example covered in detail (Nayaka of India, Bird-David 2017; Martu of Australia, Bird et al. 2019), which they compare to a larger sample of other foragers known ethnographically. Although workshop attendees ultimately found that Bird et al. and Bird-David's scalar arguments were more similar than different, the question still remains, did either study include a sufficiently large and representative sample of foragers to make conclusions about "hunter-gatherer" social scales?

Acknowledgements

Some of the research described herein was funded by a grant from the National Science Foundation of the United States of America, BCS 1733917. Special thanks to Dr. Tsiazonera, Dr. Jaovola Tombo, Patricia Hajasoa, Soanahary Gérard, Rolland Lahiniriko, Angelah Halatiana Garçon, Gervais Tantely, Théodore Tsitindry Ramanovontsoa, Jean-Claude Alhayess, Repapa Pamphil de la Patience, and Eric Rambeloson, and Eric Ringen.

References

Astuti, Rita. 1995. 'The Vezo are not a kind of people': Identity, difference, and 'ethnicity' among a fishing people of western Madagascar. *American Ethnologist* 22(3): 464-482.

Astuti, Rita, Gregg E. A. Solomon, and Susan Carey. 2004. *Constraints on Conceptual Development: The Case of the Acquisition of Folkbiological and Folksociological Knowledge in Madagascar*. Boston: Blackwell.

Barley, Shanta. 2012. Last Stand of the Madagascan Spider Tortoise. http://www.bbc.co.uk/nature/19161652 (accessed September 3, 2012).

Barnard, Alan. 1999. Images of hunter-gatherers in European social thought. In *The Cambridge Encyclopedia of Hunters and Gatherers*. Richard B. Lee and Richard Daly, (eds.). Pp. 375-383. Cambridge: Cambridge University Press.

Barth, Frederick. 1969a. Introduction. In *Ethnic Groups and boundaries: The social organization of cultural difference*. Frederick Barth, ed. Pp. 9-38. Bergen: Universitet Forlaget.

— 1969b. *Ethnic Groups and Boundaries: The Social Organization of Cultural Difference*. Bergen: Universitet Forlaget.

Bird, Douglas W., Rebecca Bliege Bird, Brian F. Codding, and David W. Zeanah. 2019. Variability in the organization and size of hunter-gatherer groups: Foragers do not live in small-scale societies. *Journal of Human Evolution* 131: 96-108.

Bird-David, Nurit. 2017. Before Nation: Scale-Blind Anthropology and Foragers' Worlds of Relatives. *Current Anthropology* 58(2): 209-226.

Borgerhoff Mulder, Monique, Samuel Bowles, Tom Hertz, Adrian Bell, Jan Beise, Greg Clark, Ila Fazzio, Michael Gurven, Kim Hill, Paul L. Hooper,

William Irons, Hillard Kaplan, Donna Leonetti, Bobbi Low, Frank Marlowe, Richard McElreath, Suresh Naidu, David Nolin, Patrizio Piraino, Rob Quinlan., Eric Schniter, Rebecca Sear, Mary Shenk, Eric Alden Smith, Christopher von Reuden, and Polly Wiessner. 2009. Intergenerational wealth transmission and the dynamics of social inequality in small-scale societies. *Science* 326: 682-688.

Buechler, Hans C., and Judith-Maria Buechler. 1970. *Bolivian Aymara.* New York: Holt, Rinehart, Winston.

Chagnon, Napoleon. 2012. *The Yanomamö*, 6th edition. Boston: Cengage Learning.

Choi, Jung-Kyoo, and Samuel Bowles. 2007. The coevolution of parochial altruism and war. *Science* 318(636): 636-640.

Crocker, William H., and Jean G. Crocker. 2004. *The Canela: Kinship, Ritual, and Sex in an Amazonian Tribe, 2nd ed.* Belmont, CA: Wadsworth/Thomson Learning.

Dewar, Robert E., Chantal Radimilahy, Henry T. Wright, Zenobia Jacobse, Gwendolyn O. Kelly, and Francesco Bernag. 2013. Stone tools and foraging in northern Madagascar challenge Holocene extinction models. *Proceedings of the National Academy of Science* 110(31): 12583-12588.

Dina, Jeanne, and Jean-Michel Hoerner. 1976. Etude sur les Populations Mikea du Sud-Ouest de Madagascar. *Omaly sy Anio* 3-4: 269-286.

Dockrill, Peter. 2020. Hunter-gatherers sit as much as us, but how they sit makes all the difference. https://www.sciencealert.com/hunter-gatherers-are-as-sedentary-as-us-but-how-they-sit-makes-all-the-difference (accessed July 1, 2021).

Drury, Robert. 1826 [1729]. *The Pleasant and Surprising Adventures of Robert Drury during His Fifteen Years' Captivity on the Island of Madagascar*. London: Hunt and Clarke.

Durkheim, Emile. 1982[1895]. *The Rules of Sociological Method.* W. D. Halls, trans. New York: The Free Press.

Evans-Pritchard, E. E. 1940. *The Nuer: A Description of the Modes of Livelihood and Political Institutions of a Nilotic People*. Oxford: Oxford University Press.

Eggert, Karl. 1986. Mahafaly as misnomer. In *Madagascar: Society and History.* Conrad P. Kottak, Jean-Aimé Rakotoarisoa, Aidan Southall, and Pierre Vérin, (eds.). Pp. 321-353. Durham: Carolina Academic Press.

Ember, Carol R. 1978. Myths about hunter-gatherers. *Ethnology* 17: 439-448.

García, Julián and Jeroen C. J. M. van den Bergh. 2011. Evolution of parochial altruism by multilevel selection. *Evolution and Human Behavior* 32: 277-287.

Gourevitch, Philip. 1998. *We Wish To Inform You That Tomorrow We Will Be Killed With Our Families: Stories From Rwanda.* New York: Farrar, Straus, and Giroux.

Hamilton, Marcus J., Bruce T. Milne, Robert S. Walker, Oskar Burger, and James H. Brown. 2007. The complex structure of hunter-gatherer social networks. *Proceedings of the Royal Society, B.* 274: 2195-2202.

Handley, Carla and Sarah Mathew. 2020. Human large-scale cooperation as a produce of competition between cultural groups. *Nature Communications* 11: 702.

Hart, C. W. M., and Arnold Pilling. 1963. *The Tiwi of North Australia.* New York: Holt, Rinehart and Winston.

Henrich, Joseph, Robert Boyd, Samuel Bowles, Colin Camerer, Ernst Fehr, Herbert Gintis, Richard McElreath, Michale Alvard, Abigail Barr, Jean Ensminger, Natalie S. Henrich, Kim Hill, Francisco Gil-White, Michael Gurven, Frank Marlowe, John Q. Patton, and David Tracer. 2005. Economic man in cross-cultural perspective: Behavioral experiments in 15 small-scale societies. *Behavioral and Brain Sciences* 28(6): 795-855.

Hill, Kim R., Robert S. Walker, Miran Božičević, James Eder, Thomas Headland, Barry Hewlett, A. Magdalena Hurtado, Frank Marlowe, Polly Wiessner, and Brian Wood. 2011. Co-residence patterns in hunter-gatherer societies show unique human social structure. *Science* 331(6022): 1286-1289.

Hodgson Dorothy L. 1999. Pastoralism, patriarchy and history: Changing gender relations among Maasai in Tanganyika, 1890-1940. *The Journal of African History* 40(1): 41-65.

Huff, Amber R. 2012. Exploring discourses of indigeneity and rurality in Mikea Forest environmental governance. *Madagascar Conservation and Development* 9(S2): 58-69.

Iliffe, John. 1979. *A Modern History of Tanganyika.* Cambridge: University of Cambridge Press.

Jones, Doug. 2018. Kin selection and ethnic group selection. *Evolution and Human Behavior* 39: 9-18.

Kelly, Robert L., 2013. *The Lifeways of Hunter-Gatherers: The Foraging Spectrum.* Cambridge: Cambridge University Press.

Kent, Raymond K. 1970. *Early Kingdoms in Madagascar 1500-1700.* New York: Holt, Rinehart, and Winston.

Kull, Christian. 2000. Deforestation, erosion, and fire: Degradation myths in the environmental history of Madagascar. *Environment and History* 6(4): 423-50.

Lamphear, John. 1988. The people of the grey bull: The origin and expansion of the Turkana. *Journal of African History* 29: 27-39.

Lee, Richard B., and Richard Daly. 1999. Introduction: Foragers and others. In *The Cambridge Encyclopedia of Hunters and Gatherers*. Richard B. Lee and Richard Daly, (eds.). Pp. 1-21. Cambridge: Cambridge University Press.

Linnekin, Jocelyn, and Lin Poyer. 1990. *Cultural Identity and Ethnicity in the Pacific*. Honolulu: University of Hawaii.

Linton, Ralph. 1928. Culture areas in Madagascar. *American Anthropologist* 30(3): 363-390.

Mack, John. 1986. *Madagascar, Island of the Ancestors*. London: British Museum.

Mafeje, Archie. 1971. The ideology of 'tribalism'. *The Journal of Modern African Studies* 9(2): 253-261.

Majid, Asifa, and Nicole Kruspe. 2018. Hunter-gatherer olfaction is special. *Current Biology* 28(3), 409-413.

Mathew, Sarah, and Robert Boyd. 2011. Punishment sustains large-scale cooperation in prestate warfare. *Proceedings of the National Academy of Sciences* 108(28): 11375-11380.

McCabe, J. Terrence. 2004. *Cattle bring us to our enemies: Turkana ecology, politics, and raiding in a disequilibrium system*. Ann Arbor: University of Michigan Press.

McElreath, Richard, Robert Boyd, and Peter J. Richerson. 2003. Shared norms and the evolution of ethnic markers. *Current Anthropology* 44(1): 122-129.

Middelton, Karen. 1999. Introduction. In K. Middleton (ed.). Pp. 1-36. *Ancestors, Power, and History in Madagascar*. Leiden: Brill.

Molet, Louis. 1958. Aperçu sur un groupe nomade de la forêt épineuse des Mikea. *Bulletin de l'Academie Malgache* 36: 241-s43.

— 1966. Les Mikea de Madagascar: Où vivre sans boire. *Revue de Madagascar* 36: 11-16.

Mouyon, Jean-Claude, and Patrice Francelle. 1999. Feu Mikea. *L'Express* (Antananarivo), June 29.

Moya, Cristina, and Robert Boyd. 2015. Different selection pressures give rise to distinct ethnic phenomena: A functionalist framework with illustrations from the Peruvian Altiplano. *Human Nature* 26: 1-27.

Murdock, George P. 1967. Ethnographic atlas: A summary. *Ethnology* 6(2): 109-236.

Oba, Gufu. 2011. Colonial resource capture: triggers of ethnic conflicts in the Northern Frontier District of Kenya, 1903-1930s. *Journal of Eastern African Studies* 5(3): 505-534.

Pierron, Denis, Harilanto, Razafindrazaka, Luca Pagani, François-Xavier Ricaut, Tiago Antao, Melanie Capredon, Clément Sambo, Chantal Radimilahy, Jean-Aimé Rakotoarisoa, Roger M. Blench, Thierry Letellier, and Toomas Kivisild. 2014. Genome-wide evidence of Austronesian-Bantu admixture and cultural reversion in a hunter-gatherer group of Madagascar. *Proceedings of the National Academy of Sciences* 111(3): 936-941.

Poyer, Lin, and Robert L. Kelly. 2000. Mystification of the Mikea: Constructions of Foraging Identity in Southwest Madagascar. *Journal of Anthropological Research* 56: 163-185.

Raichlin, David A., Herman Pontzer, Theodore W. Zderic, Jacob A. Harris, Audax Z. P. Mabulla, Marc T. Hamilton, and Brian Wood. 2020. Sitting, squatting, and the evolutionary biology of human inactivity. *Proceedings of the National Academy of Sciences* 117(13): 7115-7121.

Ranger, Terence. 1993. The invention of tradition revisited: the case of colonial Africa. In *Legitimacy and the State in Twentieth-century Africa*. Terence Ranger and Olufemi Vaughan, (eds.). Pp. 62-111. London: Palgrave Macmillan.

Rarojo, Jeannot. 1999. Ny Mikea: tsy lolo. [Les gens de la forêt] *Vintsy* 26: 4-7.

Razafindrazaka, Harilanto, François-Xavier Ricaut, Murray P. Cox, M. P., Maru Mormina, Jean-Michel Dugoujon, Louis P. Randriamarolaza, Evelyne Guitard, Laurie Tonasso, Bertrand Ludes, and Eric Crubézy. 2010. Complete mitochondrial DNA sequences provide new insights into the Polynesian motif and the peopling of Madagascar. *European Journal of Human Genetics* 18: 575-581.

Repoblikan'i Madagasikara. 2010. *Plan de Développement de la Population Autochtone Mikea*. Antananarivo: Ministère de l'Environnement, des Eaux et Forêts/Madagascar National Parks.

Richerson, Peter J., and Robert Boyd. 2005. *Not by Genes Alone: How Culture Transformed Human Evolution*. Chicago: University of Chicago Press.

Riolo, Rick L., Michael D. Cohen, and Robert Axelrod. 2001. Evolution of cooperation without reciprocity. *Nature* 414: 441-443.

Scales, Ivan R. 2012. Lost in Translation: Conflicting Views of Deforestation, Land Use and Identity in Western Madagascar. *The Geographic Journal* 178(1): 67-79.

Schlee, Günther. 1990. *Identities on the Move: Clanship and Pastoralism in Northern Kenya*. Manchester: Manchester University Press.

Schrire, Carmel. 1984 Wild surmises on savage thoughts. In *Past and Present in Hunter Gatherer Studies*. Carmel Schrire (ed.). Pp. 2-25. Orlando: Academic Press, Inc.

Science Daily. 2021. Hunter-gatherers agree on what is moral, but not who is moral. https://www.sciencedaily.com/releases/2019/09/190905080039.htm (accessed October 29, 2021).

Simpson, George L., Jr., and Peter Waweru. 2021. The implausible persistence of pastoralism: Samburu transhumance from their nineteenth-century origins through the period of colonial rule. *The Journal of the Middle East and Africa* 12(2): 225-249.

Smith, Daniel, Philip Schlaepfer, Katie Major, Mark Dyble, Abigail E. Page, James Thompson, Nikhil Chaudhary, Gul Deniz Salali, Ruth Mace, Leonora Astete, Marilyn Ngales, Lucio Vinicius, and Andrea Bamberg Migliano. 2019. Cooperation and the evolution of hunter-gatherer storytelling. *Nature communications* 8(1): 1-9.

Smith, Eric Alden, Kim Hill, Frank W. Marlowe, David Nolin, Polly Wiessner, Michael Gurven, Samuel Bowles, Monique Borgerhoff Mulder, Tom Hertz, and Andrian Bell. 2010. Wealth transmission and inequality among hunter-gatherers. *Current Anthropology* 52(1): 19-34.

Smith, Kristopher M., and Coren L. Apicella. 2020. Hadza hunter-gatherers disagree on perceptions of moral character. *Social Psychology* 11(5): 616-625.

Southall, Aidan W. 1970. The illusion of tribe. In *The Passing of Tribal Man in Africa*. P. C. W. Gutkind (ed.) Pp. 28-50. Leiden: Brill.

— 1971. Ideology and group composition in Madagascar. *American Anthropologist* 73(1): 144-164.

Spector, Tim. 2017. What a hunter-gatherer diet does to the body in just three days. CNN, July 5, 2017. https://www.cnn.com/2017/07/05/health/hunter-gatherer-diet-tanzania-the-conversation/index.html

Tucker, Bram. 2003. Mikea origins: Relicts or refugees? *Michigan Discussions in Anthropology* 14: 93-215.

— 2019. Ecological, historical, and social explanations for low rates of food sharing among Mikea foragers of Southwest Madagascar. In *Towards a Broader View of Hunter-Gatherer Sharing*. David Friesem and Noa Lavi, (eds.). Pp. 237-250. Cambridge: McDonald Institute for Archaeological Research, University of Cambridge.

— 2020. *Où vivre sans boire* revisited: Water and political-economic change among Mikea hunter-gatherers of southwestern Madagascar. *Economic Anthropology* 7(1): 22-37.

Tucker, Bram, Erik J. Ringen, Tsiazonera, Jaovola Tombo, Patricia Hajasoa, Soanahary Gérard, Rolland Lahinirko, and Angelah Halatiana Garçon. 2021. Ethnic markers without ethnic conflict: Why do interdependent

Masikoro, Mikea, and Vezo of Madagascar signal their ethnic differences? *Human Nature* 32(3): 529-556.

Yount, James W., Tsiazonera, and Bram Tucker. 2001. Constructing Mikea identity: Past or present links to forest and foraging. Ethnohistory 48: 3257-3291.

Comment by Thomas Widlok

Every discipline has its default connotations when the notion of "scale" is being invoked. Maybe most intensively this is being debated in geography where scale is being delimited by other spatial concepts such as such as place, locality, territory and space. For anthropology with its interest in socio-cultural scale rather than geographical scale the key association that features most prominently is that of 'ethnic identity'. This is so even though many anthropologists today will be quick to claim that they are not privileging ethnic groups in their research since they include a host of other, different groupings ranging from professional or age groups to more diffuse entities such as milieus, situations and subcultures. 'Ethno'graphy, too, the disciplinary method-books underline (see Breidenstein et al. 2013: 32), is today often not about ethnic groups, and there is no immediate reason why it should be. All these qualifications notwithstanding it is important to seek to clarify the relation between scale and the notion of ethnic group, as Bram Tucker does in this article. Because there is a latent danger in anthropological writing that the ethnic group may be assumed to be the default (if not the 'natural') scale of anthropological description and analysis. There are a number of reasons for these latent slippages that constitute errors of scale. One is that a good part of the body of anthropological literature up to this point has been framed in this way, not only as a habit of speaking amongst authors but also due to influential book series such as the 'Case Studies in Cultural Anthropology' and due to influential database projects such as the 'Human Relation Area Files'. Another reason has to do with the dominant mode of (lateral) comparison in anthropology and archaeology which conceives of case studies as datapoints that are commonly given ethnic labels. This is still common practice even when we do not know whether that label was used as a self-identification back in time, e.g. in much of the archaeological record. In the ethnographic record, too, it is important to take sufficient precautions in order to avoid errors of scale that 'overattribute' ethnic identity with social practices. After all, ethnic identity and ethnic identification are political resources employed for a range of purposes (see University of Cologne Forum 2015). Tucker's contribution clearly distinguishes two forms of overattribution: Cultural practices (or traits) may be ascribed to the scale of ethnic groups even though they actually correspond to larger units such as languages or regions (Type 1 error) or despite the fact that they are actually tied to smaller units such as gender groups or individuals (Type 2 error).

What is important to note in this context is that these are not purely academic concerns. As Tucker shows with his examples these errors have very direct political consequences and they do not affect everyone in the same way: Small indigenous groups more often than others suffer discrimination as a direct result of being wrongly described in the Typ 1 erroneous mode ('Mikea hunting endangers tortoises'). Hunter-gatherer groups experience also a larger than average share of Type 2 erroneous misrepresentation because being included in the category 'hunter-gatherer' often goes with assumptions of a stable evolutionary stage of early humanity ('Foragers as early humans share more food than others') and it belies the diversity found among hunter-gatherers. Tucker's comparison of two case studies suggest that ethnic groups may indeed at times be a relevant scale to consider, but not necessarily so. There is no short-cut that would spare us the trouble of carefully testing which cultural practice can be associated with what type of grouping at the various scales under consideration. Caution is required, and Tucker provides some hands-on recommendations of how to practice this caution in scholarly writing. One of these recommendations is to refer to the group's own self-designation. And this may be one of the most relevant insights here: We scale as scholars, and we need to note the likely errors that occur when we do so. But we are also always constantly dealing with people who themselves are involved in scaling as a practice. These are first and foremost the interlocutors in our field research (or for the archaeological record those who leave marks of cultural distinction in materials and landscapes). But it also involves the readers of scholarly work who employ the scales that are inbuilt into their own biases, including those that have sedimented from previous scientific work that scholars today have come to criticize and reject. The bad news is that scaling is ongoing and it is a situated practice, a response to particular contexts so that there will not be a one-scale-fits-all. The good news is that if the practice of scaling is malleable, we do stand a chance to positively influence the harmful and erroneous scaling that we observe.

References

Breidenstein, Georg, Stefan Hirschauer, Herbert Kalthoff, and Boris Nieswand. 2013. *Ethnografie. Die Praxis der Feldforschung*. Konstanz: UVK.

University of Cologne Forum Ethnicity as a Political Resource (ed.) 2015. *Ethnicity as a Political Resource. Conceptualizations across Disciplines, Regions, and Periods*. Bielefeld: transcript.

Scaling an island of hunter-gatherers
Writing the Mesolithic of Ireland

Graeme Warren

Introduction

In Spring 2020 I was nearing the end of a four-year term as Head of the UCD School of Archaeology and looking forward to the sabbatical that would follow. One of the main aims of the sabbatical was to write a long-planned book on the Mesolithic of Ireland (c. 8000-4000 BC) – *Hunter-Gatherer Ireland: making connections in an island world* (Warren 2022). Each chapter was to start with a detailed account of an artefact, based on proposed work in museums and archives. I was also to travel internationally and help develop comparative approaches to the Irish material. With lockdown imposed from March 2020, none of this could happen. My professional horizons shrank to a desk in the corner of a bedroom, with occasional Zoom connections to other places. My social world revolved almost exclusively around my wife, two children and our cats. And of course, the book changed as the scales at which I could operate changed. The book reflected my increasing, desk-bound, interest in how we could make statements about hunter-gatherer lives in the deep-time past, how such knowledge claims worked and what value such narratives might have during a time of crisis. The latter included reflection on the ways in which archaeological accounts of long dead hunter-gatherers might be relevant to those with an interest in the lives of contemporary hunter-gatherers.

As I began to prepare the book, two short statements about the Mesolithic in Ireland made in recent publications were very much in my mind. These comments were by two senior Irish archaeologists, including one who would have been regarded as the leading Mesolithic researcher in Ireland, and they suggested that aside from a small number of spectacular sites, the large number of archaeological excavations conducted in advance of commercial or infrastructural development in Ireland had contributed very little to our understanding of hunter-gatherer activity in Ireland (Waddell 2020: 54; Woodman 2015: 79), and therefore made little contribution to broader understandings of hunter-gatherer lives. These statements frustrated and angered me. Dismissing the results of the single largest phase of archaeological work ever seen in the history of Ireland as having made little contribution to how we understood deep-time hunter-gatherers did not seem to place value on the work of many colleagues across the profession who had painstakingly excavated Mesolithic sites, sometimes in very difficult conditions. Devaluing these contributions did not seem collegial and did not seem likely to encourage the careful excavation of ephemeral materials in the future.

But more importantly, these dismissals also suggested a fundamental misunderstanding of the archaeological record for the period. Many Mesolithic sites excavated in advance of infrastructural development in Ireland are characterised by small spreads of archaeological material, scattered hearths, and a few stone tools. Structural evidence is rare. These sites are *not* spectacular. But to dismiss them as not contributing to our understanding of the period is wholly to miss the point. These sites are the dominant form of Mesolithic archaeology in Ireland. They have been demonstrated to be such by the large-scale excavations that could only have been carried out in archaeological interventions in advance of development. Understanding the Mesolithic of Ireland therefore means foregrounding this material and the activities of hunter-gatherers in the past that generated these sites. One of the aims of my book became to create a narrative that engaged this material and provided a framework for such evidence.

The opportunity to attend the *Scale Matters* workshop helped refine my growing realisation that many of my concerns about how we could make statements about the past and the value of those statements were resolved by careful consideration of scale. In particular, rethinking the material recovered during infrastructural works and dismissed by other accounts means rethinking what this material tells us about scale: including the scales at which archaeological evidence is resolvable, and the scales at which lives were lived in the past.

The remainder of this chapter therefore considers three aspects of how scale articulates my approach to the Irish Mesolithic, and provides a case study for the importance of scale, and associated concerns about analytical resolution, in writing narratives about deep-time hunter-gatherers in other places. In turn, we will consider temporal scale; social scale; and finally, and in brief, how we might quantify scale. Many of the themes discussed in brief here are considered in substance – and with fuller references and examples – in *Hunter-Gatherer Ireland*.

Temporal scale

It is often stated that archaeology's distinctive strength is the long-term perspective that it provides on human lives. It is true that much of our data is resolvable only with comparatively coarse chronological resolution. Time perspectivism stressed that archaeology needed to tailor its questions to the temporal resolution of its data (Bailey 2007, 2008) and questioned whether archaeology should be trying to apply concepts drawn from disciplines characterised by different temporal resolution – such as social or cultural anthropology. At times, strong versions of this position imply that archaeology can *only* provide data on the long-term (Perreault 2019; Kelly this volume): in such accounts our strength is also our limit.

I find this conclusion regarding the potential temporal scale of hunter-gatherer archaeology to be unduly pessimistic. I also think that it does not adequately characterise the nature of the archaeological record of the Mesolithic (see also Elliott and Griffiths 2018). Negotiating the evidence of hunter-gatherer lives in the deep-time Irish past requires engaging with data which has very different chronological resolution and using these different scales of analysis to highlight key aspects of that data. Three 'types' of chronological scale are reviewed here: long term and persistent places; places used only for short periods of time, and individual moments – or perhaps 'situations', All of my examples lie within the Later Mesolithic of Ireland – and with all the examples cited falling broadly within the period c 6000-4000 BC.

We start with the long-term and the coarse temporal resolution. A key feature of the Mesolithic landscape at this time was locations that were repeatedly returned to over the long term. Recent excavations at coastal sites like Ferriter's Cove, Co. Kerry (Woodman, Anderson, and Finlay 1999); Belderrig, Co. Mayo (Warren 2009) and Fanore, Co. Clare (Lynch 2017) show that individual episodes of occupation left little clear structural trace – occasional pits, hearths or stakeholes – but that visits to these locations took place over periods of hundreds of years. We do not know if these visits took place every year, or whether there were gaps in otherwise continuous cycles of occupation. In any case, in some of these places, this resulted in truly time-averaged archaeological deposits: 'occupation soils' containing the accumulated and mixed materials of multiple individual acts. Disaggregating those individual activities is not possible. Analysis must therefore seek to explain them as long-term phenomena. These recent excavations are broadly paralleled by poorly understood mid-Twentieth Century excavations of coastal sites in eastern Ireland: because they are a frequently occurring site type, understanding these long-term sites is crucial to considering the character of Later Mesolithic activity in Ireland.

Many of these sites were associated with the exploitation of local raw materials for stone tools, and the large accumulations of so-called waste materials which were found in the excavations would have been observable in the past. They may have acted as material prompts and traces of previous activities, as well as convenient places to find lithic material to use as tools without the need for further flaking (Dibble et al. 2017: 829). Seasonal evidence from these persistent coastal places suggests presence in autumn, at least, and in-shore marine fishing appears common, as well as patchy evidence for exploitation of terrestrial plants and animals.

Whatever the specifics about activity on individual sites these places were frequently visited at time scales beyond those of individual life spans. They were *persistent places* in the Mesolithic landscape. As Shaw and colleagues have argued

> "Places that groups return to repeatedly are invested with the qualities of the interactions that have taken place before—whether they are held in direct memory, or inferred from observable traces (old fireplaces, reused lithics, bone refuse). A persistent place possesses different qualities as a locale ... to a transient camp because it is overlain with this enhanced patina of extended social life." (Shaw et al. 2016: 1450).

Temporal scale, as experienced in the past, created different senses and experiences of place. Understanding our long-term sites should recognise that they created particular experiences of time, whatever precise form this recognition of previous activity took.

So far, so long-term. These persistent places were often excavated in a research-led context: with researchers drawn to highly visible accumulations of shell or frequently discarded local lithic raw materials. Ironically, the poor chronological resolution of the sites enabled the greater archaeological visibility and greater research interest. Given this high archaeological visibility, it is therefore a little surprising that such sites are quite rare in development-led archaeology. Instead, the large areas excavated on infrastructural projects have shown the importance of small sites, sometimes without any associated artefacts. At Curraghprevin 3, Co. Cork, for example, a fire-setting and few stake holes was argued to be a short-term activity area (Hanley and Hurley 2013). Excavations at Farriters, Co. Tyrone found an isolated hearth dating to the mid fifth millennium BC (Site 35) and a pit from a few centuries later (Site 36) (Dunlop and Barkley 2016). At Tinryland 1, Co. Carlow, excavations revealed a hollow dating to about 4000 BC with a few Later Mesolithic stone tools and carbonised hazelnuts. Although it is not stated explicitly, sites of these kinds were probably in the minds of Waddell and Woodman when they made their comments on the limited contribution of developer led archaeology. And of course, these small sites are unlikely to have attracted the attention of research-led excavations, partly because they often appear on complex multi-period sites, but also because they would be very hard to find and not necessarily repay the limited resources and need for impressive results of research led excavation (for discussion of the value of 'small sites' see Marchand and Goffic 2009)

These short-lived places provide another perspective on temporal scale and therefore on hunter-gatherer lives in Ireland. The resolution of our radiocarbon dates means that we can only date these events to some point within a few hundred years, but in most instances the activities that gave rise to the archaeological evidence need only have taken a few hours or days. One interesting feature of these places is that they were not revisited over long periods of time. They did not become *persistent*. Activity in these places may therefore have lacked the 'patina' of previous activity enabled by more persistent locales, and they speak to us of different experience in the past. It is not clear *why* these locations did not develop and persist. But making sense of hunter-gatherer use of landscapes requires that we consider these shorter-term visits, as well as the long-term aggregates.

Finally, our archaeological evidence includes individual moments. These are often hard to access – such as the blows of a stone hammer that removed a flake from a core; or the dump of material into a pit. But sometimes those moments are vivid. A good example is in the presence of pine tapers on sites such as Corralanna (Warren, Little, and Stanley 2009), Derragh (Fredengren 2009) and Moynagh Lough (Bradley 2001), with slightly different examples from Clowanstown (Mossop and Mossop 2009). These tapers are short lengths of wood with charring at one or both ends. They are most likely to have been used as sources of light, with the resinous pine wood chosen for the quality of its flame. They were transient artefacts – consumed in the act of their use.

We can't be certain what they were used for, their frequent recovery on lake-edge platforms suggests a role in using light as a lure for fishing, but as they would only be preserved on water-logged sites, there is a circularity in this argument. Perhaps they were torches to light a journey, or to accompany dance, song, or other rites.

Whatever their specific use, consideration of the tapers has the potential to provide a connection with short term processes and activities in the hunter-gatherer past. This connection is partly analytical and partly empathetic: with the imagined use of an object and the places and social contexts that it illuminated providing a powerful point of engagement with the past. These types of connections resonate with the hunter-gatherer situations described by Widlok: "the social space created by particular practices that are associated with hunter-gatherer ways of life" (Widlok 2016a). The short-term use of flickering torches is an example of the way in which hunter-gatherer situations allow us to engage with the deep time past. The creation of persistent places, and the experience of the patina of previous lives was another hunter-gatherer situation in Ireland: a practice that was associated with their way of life and generated senses of time and place.

The temporal scales that characterise the evidence of hunter-gatherer lives in Ireland therefore range from aggregates across centuries and millennia to moments that lasted minutes and hours – even if we can't always say exactly when these moments took place. The craft of writing a narrative of hunter-gatherer lives in Ireland means moving between these scales and using each of them to illuminate the others. We can play to our long-term strengths, but also highlight moments of contact and connection. To emphasise one temporal scale at the expense of the others would be a loss.

Scales of social life

A key assumption that operationalises the analysis in *Hunter-Gatherer Ireland* is that much of the evidence from Ireland is in keeping with hunter-gatherers who had a reasonably high degree of routine residential mobility and relatively small residential group size. This assumption is drawn, inductively, from twenty years of my work on the period. It is an assumption that can, and should, be questioned by others. But the value in making this assumption explicit in my analysis is the access to comparative and general models of hunter-gatherer behaviours that it enables.

This included two key areas of work on hunter-gatherer sociality that had inspired me. On the one hand, many recent accounts emphasise the importance of the intensity of intimately shared presence in hunter-gatherer groups and the ways in which this is central to key aspects of hunter-gatherer sociality (amongst many, see Bird-David 2017a; 2017b; Hewlett et al. 2019; Widlok 2016b). On the other hand, was the strong statement that "foragers do not live in small scale societies" (Bird et al. 2019). Coincidentally, these two seemingly contradictory considerations of scale were a key point of discussion in the *Scale Matters* workshop. Simplifying crudely, these discussions implied that given conditions of high mobility and small group size I could assume that sociality within a band would be characterised by a high degree of intensity and flexibility associated with varied acts and types of sharing; but that contact between groups over long distances should also exist. I found this assumption about scales and their implications very helpful in thinking through two aspects of the Irish data: firstly, the absence of evidence for structures on many sites and secondly, the evidence for contact over distance.

Evidence for what might be considered domestic buildings or structures on Irish Mesolithic sites is comparatively rare, especially for the Later Mesolithic. Most accounts stress that issues of taphonomy and loss have been significant in shaping the record: arguing that soil formation processes or later phases of activity have disturbed or removed the evidence of Mesolithic buildings. This emphasis on the role of taphonomy is important – but beyond this, there is also a lack of precision in terminology and analysis: poorly defined 'huts', 'shelters', 'wind-breaks' are described as constituting 'campsites'. The lack of precision, and the assumptions of taphonomic loss, mean that there has been little attempt to make sense of the nature of domestic architecture.

An analytical framework originally applied to Norwegian Mesolithic structures (Fretheim et al. 2018; Fretheim 2017) defines tents as 'portable dwellings, built to be easily assembled, disassembled and transported' and stresses that in many instances they leave very little clear archaeological trace in terms of structural evidence, precisely because they are designed to be mobile. Communities reliant on tents may therefore leave little direct evidence of those structures, with sites dominated by scattered artefacts and isolated features. Such a description is in keeping with much of the Irish evidence of accumulated occupation soils, spreads and occasional pits and fire settings. We have occasional (semi-)permanent buildings and some 'composite buildings' (where some structural features such as poles might be left on site for re-use), but tents appear to have been more common, especially in the Later Mesolithic. The precise form of the structures is probably not identifiable, but the choice appears to have been significant in shaping the archaeological record.

But why choose to live in tents? Unfortunately, beyond the assumption that mobility was essential to Mesolithic lives, this question is rarely considered in the Irish literature – not least because of the absence of terminological clarity and the pervasive power of stereotypes of hunter-gatherer behaviour. Drawing on the work on hunter-gatherer social scale outlined above, my argument is that the reliance on tents was a choice made by deep-time hunter-gatherers in Ireland *because* living in tents enabled the proximity and intimacy so important to their social worlds. As Friesem and Lavi observe

> "the rule of thumb among hunting and gathering societies is that houses are open or semi-open structures, built with very light and easily modified materials. Above all, the house design and site structure among foragers seems to manifest a social preference to ensure maximum sharing, co-presence and living-together." (Friesem and Lavi 2019: 88-9)

Choosing to live in tents was an option that asserted and maintained a form of hunter-gatherer sociality in Mesolithic Ireland, a form of intense and shared presence characteristic of living in small social groups. The intimate scales of hunter-gatherer social life thus give rise to our evidence. Rather than simply representing taphonomic loss, therefore, the absence of structural evidence on many Mesolithic sites in Ireland can be interpreted as resulting from choices made in the past to emphasise the creation of intimate co-presence, itself enabling the maintenance of key institutions such as sharing. These choices arguably arose from an emphasis on forms of relational wealth, and, presumably, acted to inhibit the inheritance of material wealth or power. Reconstruction drawings commissioned for the volume try and capture something of these moments (Figure 1). Some evidence suggests possible population growth in the final phases of the Irish Mesolithic (Chapple, McLaughlin, and Warren accepted) and it is interesting to note that this increase does not appear to be associated with the significant development of social inequality many general evolutionary models would suggest: perhaps the strong assertion of intimate living was a way of resisting this?

Alongside the creation of intimate spaces for small-scale sociality, Irish hunter-gatherers appear to have maintained long-distance contacts – at least across the island of Ireland. A common feature in Later Mesolithic lithic as-

Figure 1: Shared Mesolithic moments

semblages is the presence of (very) small amounts of non-local raw materials, often drawn from distances of c 100-200km. These include sites on the coast such as Bay Farm, Co Antrim (Woodman and Johnson 1996), Belderrig (Warren 2009) and Ferriter's Cove (Woodman, Anderson, and Finlay 1999) as well as inland at Clogheen, Co. Waterford (Kador 2007) and Lough Derravaragh, Co. Westmeath (Little 2010). Most of the time these are finished objects and they do not appear to be especially different in terms of their functional characteristics or possible 'prestige'. They are often only a handful, or even single, artefacts in assemblages dominated by local raw materials.

Multiple interpretations of the precise processes by which these artefacts travelled over distance are possible, but it is most parsimonious to simply assume that they represent contact of some kind between different groups or across distance. Following the observations on scales of hunter-gatherer sociality reviewed above, the contacts indicated might be considered an example of suggestion by Bird et al. of a

> "model for hunter-gatherer group formation in which fluid groups of co-residing/co-working individuals are not drawn from a small well-defined community or ethnolinguistic group, but rather from networks of social organization maintained in relational, rather than material, wealth accumulation." (Bird et al. 2019: 96)

The small amounts of 'exotic' raw materials from distance resulted from social strategies that encouraged small amounts of movement and contact between places and between groups: an emphasis on connectivity and relationships which enabled mobility.

In these examples, considering the scales of sociality characteristic of hunter-gatherers as observed ethnographically and in anthropological synthesis enables interpretation of Irish archaeological evidence not simply as taphonomic loss or a failure to find the right kinds of sites, but as resulting from the decisions of past hunter-gatherers to emphasise the generation of relational wealth arising from intimate co-presence and fluidity. Scale gives meaning and depth to our accounts.

Quantifying scale

The final reflection on how scale shapes narratives about deep-time hunter-gatherers in Ireland considers how we quantify scale. Ireland has not been isolated from the increasing popularity of demographic approaches to prehistoric social change. This has usually taken the form of statistical modelling of radiocarbon dates: in the Irish instance, different models, sometimes on limited data sets, have produced very different results for the Mesolithic (Griffiths and Robinson 2018; Riede 2009; Riede, Edinborough, and Thomas 2009; McLaughlin 2020). A variety of attempts to quantify population levels have been made, drawing on ethnographic parallels or supposed carrying capacities for different environments (Woodman 2015). Dominant narratives suggest that Ireland was isolated and with a low population level. These archaeological discussions parallel recent genomic data from two Mesolithic individuals which has also been used to argue for both a low overall population level, possibly with a significant bottleneck (Cassidy et al. 2020). Against this expectation of a limited population size Cassidy comments that the absence of evidence of inbreeding in the genomic data "is remarkable, given that the

Mesolithic population of the island is typically estimated as no more than 3,000-10,000 people" (Cassidy 2020: 34).

Such accounts attempt to provide a meaningful scale for hunter-gatherer Ireland through consideration of population, and especially the repeated emphasis on low population levels. Quantifying the scale of the past in this sense means counting *people*: scale is population and a population number seems to provide something solid to hold onto. In this sense it is interesting that a common query from members of the public is to ask what the population was at varied times in prehistory: quantification provides some comfort in the face of the unknown. But population estimates for hunter-gatherer Ireland on the basis of a wide range of general models of carrying capacity for appropriate environments and/or more-or-less explicit analogies with hunter-gatherers in similar environments provide a bewildering range of estimates (for discussion see Warren 2015). If it is comfort that is sought through their use, then at best they offer false comfort.

The specific background to Cassidy's comment about a population of 3,000-10,000 is interesting to explore. This is claimed to be provided in Woodman's 2015 discussion, although he states 3,000-5,000 (and in another place, 800-8,000). Woodman's figure of 3,000-5,000 is drawn from Lourandos' (1997) summary of the potential pre-contact population of Tasmania, itself drawn from Jones' work on the diaries of George Augustus Robinson – the 'Chief Protector of Aborigines' 1839-1849. Robinson played a significant role in the resettlement of Aboriginal Tasmanian communities in the mid nineteenth century. British scientific accounts of Tasmanian society at this time were racist and complicit in colonial atrocities: not least in creating a 'myth of extinction' – the idea that Tasmanian society was in decline prior to colonial genocide. Downplaying the size of Aboriginal populations was part of this dominant narrative. Most recent historical accounts suggest estimates of 6,000-10,000 are more appropriate (Taylor 2017). Given that Tasmania is only 81% the size of Ireland, if we want to follow this logic, this might suggest a population of 7,300-12,300. But even this logic requires caution. In this attempt to quantify some of the scales at which deep time Irish hunter-gatherers lived we are embedding knowledge about hunter-gatherers which was gained under recent conditions of colonialism into the deep time past.

Most accounts of the Irish Mesolithic that discuss population assume that the meaningful scale for understanding population is that of the island of Ireland. This is unfortunate. Setting aside the complex issue of links beyond the island of Ireland, and staying with our parallel for contact-era Tasmania –

noting that these are records of societies undergoing significant violence and disruption – it is important to consider the structure of the island-wide population of Tasmania. This was organised into about 100 clans which formed nine nations and five language groups (Ryan 2012; Taylor 2017). Marriage was often within nations and nations held a variety of different relationships with their neighbours. This example suggests that quantifying scales of population at the level of the island may not be meaningful because the scales at which lives were lived were not structured at that level.

Finally, and as observed by Nurit Bird-David (this volume), quantifying the scales of population or community– and what might be small-scale or large-scale – depends to a considerable degree on who and what you include in your counting. This important observation highlights a final theme that articulated the writing of *Hunter-Gatherer Ireland*, the connections established between people and different aspects of the worlds that surrounded them, which in turn are central to considering the scales at which lives were extended. Whilst many of the details of these relationships are hard to recover, we can observe that the first Holocene settlement of the island of Ireland in the centuries surrounding 8000 BC appears to have involved the movement of hunter-gatherers and animals – with the translocation (possibly not synchronously) of wild boar, wild cat/lynx, dog and, just possibly, bear (Warren et al. 2014). This was not just a case of humans 'colonising' an island environment, but the arrival of a multi-species community: and one which probably extended beyond the mammals listed above. The ecological impact of these varied communities on the landscape of Ireland is not well understood, with too many commentators assuming that the pre-farming landscapes of Ireland were 'natural' woodlands. A more refined understanding of Early Holocene landscapes in Ireland needs to consider not that they are anthropogenically altered, but that they are the product of multi-species communities acting at different scales and bound into relationships of differing degrees of dependence. Boar, for example, have considerable influence on ecosystems, possibly enriching them through disturbance. A different example is wild cat, which was presumably less bound with human lives, but had considerable influence on the behaviour of small mammals and ground nesting birds. Understanding the scale at which hunter-gatherer lives were lived requires that we understand relationships that extend beyond the human. And in this context, the consumption by Mesolithic communities of small amounts of birds of prey, including peregrine falcon, owl and eagle, is probably best understood not as driven by calorific need but as the consumption of some kind of property of the animals in a context where the boundaries between humans and other beings were fluid. The intimate social worlds of Irish hunter-gatherers were lived in the 'pluripresence' of multiple beings (Bird-David 2017a; 2017b). In such a world, considering scale appropriately means thinking beyond the human.

Discussion

The aim of this brief essay has been to highlight how different uses of and conceptions of scale create kinds of knowledge about the deep-time hunter-gatherer past, and about how these provide value. Considering scale appropriately is a key step in making sense of the archaeological record and scale underpins most of my analysis and interpretation in *Hunter-Gatherer Ireland*. Not recognising the importance of scale is one of the reasons that some commentators have misunderstood the character and significance of evidence from the Mesolithic period in Ireland.

Although articulated most clearly in ethnographic and anthropological work, new approaches to the scales at which hunter-gatherers live enable new interpretations of deep-time hunter-gatherer lives. Temporal scale helps to create distinctive hunter-gatherer situations. In the Irish case, this allows us to consider how temporal scale affects the experience of place and how moments in the past provide points of contact and connection from the present. The importance of the latter should not be overlooked, and certainly provides one area where public interest can be engaged.

Considering the scales at which lives were lived provides an opportunity to foreground the textures of sociality in the past, with intimate co-living resulting from deliberate choices to maintain forms of architecture that enabled the development of trust and positive relationships. The value of this relational wealth was also upheld through longer journeys that bought people together. Thinking with scale in this sense allows us to understand the forms of our evidence as more than just loss, and to emphasize choices and social strategies in the past.

Attempts to quantify the scales of hunter-gatherer sociality in terms of population and demography also require careful consideration. Beyond the specific methodologies involved it is important to highlight the assumptions about appropriate scales and analogies that articulate some discussions.

Finally, taking the scales of hunter-gatherer life seriously means that we should not restrict our focus to humans. The intense connections with others, articulated so beautifully in recent ethnographies and summaries, have to be at the heart of our archaeological approaches to hunter-gatherers.

Acknowledgements

I am grateful to Thomas Widlok and the organisers of the Scale Matters workshop for the opportunity to participate and to contribute to this volume. Dores Cruz, Thomas Widlok and an anonymous referee made useful suggestions for improvement. Ben Elliott kindly read an early draft of this paper. The reconstruction drawings are by my daughter, Sadhbh Warren.

References

Bailey, Geoff. 2007. Time perspectives, palimpsests and the archaeology of time. *Journal of Anthropological Archaeology* 26 (2): 198-223.

Bailey, Geoff. 2008. Time Perspectivism: Origins and Consequences. In *Time in Archaeology: Time Perspectivism Revisited*, edited by S. Holdaway and L. Wandsnider. Salt Lake City: University of Utah Press: 13-30.

Bird, Douglas W., Rebecca Bliege Bird, Brian F. Codding, and David W. Zeanah. 2019. Variability in the organization and size of hunter-gatherer groups: Foragers do not live in small-scale societies. *Journal of Human Evolution* 131: 96-108.

Bird-David, Nurit. 2017a. *Us, Relatives: Scaling and Plural Life in a Forager World*. Oakland: University of California Press.

Bird-David, Nurit. 2017b. Before Nation: Scale-Blind Anthropology and Foragers' Worlds of Relatives. *Current Anthropology* 58 (2): 209-226.

Bradley, John. 2001. A Late Mesolithic settlement site in Eastern Ireland. In *Recent Developments in Wetland Research*, edited by B. Raftery and J. Hickey. Dublin: Seandálaíocht; Department of Archaeology, University College Dublin Monograph Series Volume 2: 299-306.

Cassidy, Lara M. 2020. Ancient DNA in Ireland: isolation. immigration and elite incest. *British Archaeology* 174 (September/October 2020): 32-41.

Cassidy, Lara M., Ros Ó Maoldúin, Thomas Kador, Ann Lynch, Carleton Jones, Peter C. Woodman, Eileen Murphy, Greer Ramsey, Marion Dowd, Alice Noonan, Ciarán Campbell, Eppie R. Jones, Valeria Mattiangeli, and Daniel G. Bradley. 2020. A dynastic elite in monumental Neolithic society. *Nature* 582 (7812): 384-388.

Chapple, Robert, Rowan McLaughlin, and Graeme M Warren (accepted). "...Where they pass their unenterprising existence...": change over time in the Mesolithic of Ireland as shown in radiocarbon dated activity. *Proceedings of the Royal Irish Academy.*

Dibble, Harold L., Simon J. Holdaway, Sam C. Lin, David R. Braun, Matthew J. Douglass, Radu Iovita, Shannon P. McPherron, Deborah I. Olszewski, and Dennis Sandgathe. 2017. Major Fallacies Surrounding Stone Artifacts and Assemblages. *Journal of Archaeological Method and Theory* 24 (3): 813-851.

Dunlop, Colin, and Jonathan Barkley. 2016. *Road to the West: The Archaeology of the A4/A5 Road Improvements Scheme from Dungannon to Ballygawley, A Road to the Past Volume* 2. Belfast: Northern Archaeological Consultancy Ltd.

Elliott, Ben, and Seren Griffiths. 2018. Living Mesolithic Time: Narratives, Chronologies and Organic Material Culture. *Journal of World Prehistory* 31: 347-365.

Fredengren, Christina. 2009. Lake Platforms at Lough Kinale – memory, reach and place: a Discovery Programme Project in the Irish Midlands. In *Mesolithic Horizons: Papers presented at the Seventh International Conference on the Mesolithic in Europe, Belfast 2005.*, edited by S. McCartan, R. Schulting, G. M. Warren and P. C. Woodman. Oxford: Oxbow: 882-887.

Fretheim, Silje E, Hein B. Bjerck, Heidi Mjelva Breivik, and Atilio Francisco Zangrando. 2018. Tent, Hut or House? A discussion on Early Mesolithic Dwellings in Light of the Site Mohalsen 2012-II, Vega, Northern Norway. In *Early Economy and Settlement in Northern Europe: Pioneering, Resource Use, Coping with Change*, edited by H. P. Blankholm. Sheffield: Equinox: 207-230.

Fretheim, Silje Elisabeth. 2017. *Mesolithic dwellings: An empirical approach to past trends and present interpretations in Norway*. Thesis for the Degree of Philosophiae Doctor: Norwegian University of Science and Technology, Trondheim.

Friesem, David E., and Noa Lavi. 2019. An ethnoarchaeological view on hunter-gatherer sharing and its archaeological implications for the use of social space. In *Towards a Broader View of Hunter-Gatherer Sharing*, edited by N. Lavi and D. E. Friesem. Cambridge: MacDonald Institute for Archaeological Research: 85-96.

Griffiths, Seren, and Erick Robinson. 2018. The 8.2 ka BP Holocene climate change event and human population resilience in northwest Atlantic Europe. *Quaternary International* 465: 251-257.

Hanley, Ken, and Maurice F Hurley, (eds.). 2013. *Generations: the archaeology of five national road schemes in Cork. Volume 1: prehistoric sites*. Dublin: NRA Scheme Monographs 13.

Hewlett, Barry S., Jean Hudson, Adam H. Boyette, and Hillary N. Fouts. 2019. Intimate living: sharing space among Aka and other hunter-gatherers. In *Towards a Broader View of Hunter-Gatherer Sharing*, edited by N. Lavi and D. E. Friesem. Cambridge: MacDonald Institute for Archaeological Research: 39-56.

Kador, Thomas. 2007. Stone age motion pictures: an objects' perspective from early prehistoric Ireland. In *Prehistoric Journeys*, edited by V. Cummings and R. Johnston. Oxford: Oxbow: 33-44.

Little, Aimée. 2010. *Tasks, temporalities and textures: reconstructing the social topography of and Irish Mesolithic Lakescape.* UCD School of Archaeology.

Lourandos, Harry. 1997. *Continent of Hunter-Gatherers: New Perspectives in Australian Prehistory*. Cambridge: Cambridge University Press.

Lynch, Michael. 2017. The Later Mesolithic on the North-West Coast of Clare. *Archaeology Ireland* 31 (4): 24-28.

Marchand, Grégor, and Michel Goffic. 2009. Give us some small sites please! Report on the first year of excavations in the rock shelter of Pont-Glas (Plouneour-Menez, Finistère, France). *Mesolithic Miscellany* 19 (2): 12-16.

McLaughlin, T. Rowan. 2020. An archaeology of Ireland for the Information Age. *Emania* 25: 7-30.

Mossop, Matt, and Emma Mossop. 2009. *M3 Clonee-North of Kells Contract 2 Dunshaughlin – Navan. Report on the Archaeological Excavation of Clowanstown 1, Co. Meath Ministerial Directions No.A008/011 E3064. Final Report.*

Perreault, Charles. 2019. *The Quality of the Archaeological Record*. Chicago: Univeristy of Chicago Press.

Riede, Felix, Kevan Edinborough, and Mark G. Thomas. 2009. Tracking Mesolithic Demography in Time and Space and its Implications for Explanations of Culture Change. In *Chronology and Evolution within the Mesolithic of North-West Europe: Proceedings of an International Meeting, Brussels, May 30th-June 1st 2007*, edited by P. Crombé, M. Van Strydonck, J. Sergant, M. Boudin and M. Bats. Newcastle: Cambrige Scholars Publishing: 177-194.

Riede, Felix. 2009. Climate and Demography in Early Prehistory: Using Calibrated 14C Dates as Population Proxies. *Human Biology* 81 (2/3): 309-337.

Ryan, Lyndall. 2012. *Tasmanian Aborigines: a history since 1803*. London: Allen & Unwin.

Shaw, Andrew, Martin Bates, Chantal Conneller, Clive Gamble, Marie-Anne Julien, John McNabb, Matt Pope, and Beccy Scott. 2016. The archaeology of persistent places: the Palaeolithic case of La Cotte de St Brelade, Jersey. *Antiquity* 90 (354): 1437-1453.

Taylor, Rebe. 2017. *Into the Heart of Tasmania: a search for human antiquity*. Melbourne: Melbourne University Press.

Waddell, John. 2020. Early Demographic Ebb and Flow in Pre-census Ireland. *Emania* 25: 53-59.

Warren, Graeme M, Stephen Davis, Meriel McClatchie, and Rob Sands. 2014. The potential role of humans in structuring the wooded landscapes of Mesolithic Ireland: a review of data and discussion of approaches. *Journal of Vegetation History and Archaeobotany* 22 (5): 629-646.

Warren, Graeme M. 2009. Belderrig: a 'new' Later Mesolithic and Neolithic landscape in northwest Ireland. In *From Bann Flakes to Bushmills; papers in honour of Professor Peter Woodman*, edited by N. Finlay, S. McCartan, N. Milner and C. J. Wickham-Jones. Oxford: Prehistoric Society Research Paper 1: 143-152.

Warren, Graeme M. 2015. Britain and Ireland inside Mesolithic Europe. In Continental connections: exploring cross-channel relationships from the Mesolithic to the Iron Age, edited by H. Anderson-Whymark, D. Garrow and F. Sturt. Oxford: Oxbow.

Warren, Graeme M. 2022. *Hunter-Gatherer Ireland: making connections in an island world*. Oxford: Oxbow: 43-58.

Warren, Graeme M., Aimee Little, and Michael Stanley. 2009. A late Mesolithic lithic scatter and radiocarbon dates from Corralanna, Co. Westmeath. *Proceedings of the Royal Irish Academy* 109C: 1-35.

Widlok, Thomas. 2016a. *Anthropology and the Economy of Sharing*. London: Routledge.

Widlok, Thomas. 2016b. Hunter-gatherer situations. *Hunter Gatherer Research* 2 (2): 127-143.

Woodman, Peter C, and Gina Johnson. 1996. Excavations at Bay Farm 1, Carnlough, Co. Antrim and the study of the 'Larnian' technology. *Proceedings of the Royal Irish Academy* 96C: 137-235.

Woodman, Peter C. 2015. *Ireland's First Settlers: Time and the Mesolithic*. Oxford: Oxbow.

Woodman, Peter C., Elizabeth Anderson, and Nyree Finlay. 1999. *Excavations at Ferriter's Cove, 1983-95: last foragers, first farmers in the Dingle Peninsula*. Bray: Wordwell.

Comment by Charlotte Damm

Archaeology is arguably challenged by the choice of scale to an even greater extent than anthropology. The discipline must deal with a variety of temporal scales as well as spatial and social ones. A fundamental issue for most archaeological projects is deciding the temporal and spatial scales most appropriate for addressing the research questions. We may wish to investigate the evolutionary development of early hominids in Africa on a continental scale and over more than a million years or identify individual flint knappers through refitting of lithic debris produced during one short knapping session extending over a few square meters – or on any scale in between.

Warren's discussion of scales provides a different angle as he contemplates social scale and temporality as perceived and experienced by past hunter-gatherers. The emphasis on the experience of scale is evident, for example, when Warren argues that small dwellings were preferred because of the intimacy they afford. An additional possibility is that tents were preferred because they could be adjusted to the number of people in any current camp.

With regard to the experience of temporality, some sites bear evidence of having been repeatedly occupied, (perhaps) a consequence of their recognisability because one had been there before, or because the place has qualities that repeatedly drew people there. The material left behind would be familiar to later inhabitants, much as we today may enter deserted ruined houses and find a broken comb or an empty matchbox. Repeated visits to such persistent places may also suggest that the location was named. The many short-lived sites, on the other hand, show us that life in the Irish Mesolithic was dominated by a rhythm of movement and a variation between new sites and persistent places. Variation is also a key element when Warren turns to social scales, where he assumes that much time was spent in small intimate groups, but that in addition interaction occurred across longer distances. Intimacy and movement were thus two central experiences in the Irish Mesolithic, but also the rhythm of variation between new and familiar sites, near and more distant regions.

What fascinates me is how Warren is first directly concerned with the resolution and scale of the *archaeological* data to be analysed, but then almost imperceptibly translates this into *lived* scales. He then takes this further, implying that the emphasis on intimacy was an inherent or perhaps even an acknowledged way of resisting social inequality.

In the final section Warren addresses the issue of who to include in our demographic counts of community members, as problematized by Bird-David. Should dogs be included? In several prehistoric hunter-gatherer contexts some dogs were buried in separate graves amongst humans in the burial grounds. Should wild animals adopted and cared for be included? And in the case of the Irish Mesolithic, several species were introduced to the island by humans, a strong indication of a multispecies community. The Mesolithic art and rock art in Europe is concerned with a limited number of species, and it is not uncommon in hunter-gatherer societies to refer to some species by kinship terms. Nurit Bird-David has previously pointed out that other-than-humans may be considered as included in a community, because what counts is the being with, rather than being like. While dealing with such issues in past hunter-gatherer communities is certainly challenging, Warren's examples of human-animal relations illustrate that the interactions go beyond a simple economic hunter-prey relationship.

Comment by Bram Tucker

Graeme Warren frames his thoughtful chapter around the experiences of pandemic lockdown in 2020, as people around the globe saw our social worlds zoom from the large scale of international travel, fieldwork, workshops, and conferences down to the small scale of living and working at home. Staying home with his wife, children, and cats forced Warren (as it did all of us) into a social world that may have been more like that of Mesolithic Irish hunter-gatherers: a social world composed of the same faces every day, a world of close kin, and of kindred extended to our non-human companions.

The other major set of events in 2020 was, in the United States at least, a public debate about exactly whose lives matter, and what it means to say that we, as a society, value the lives of Black, Brown, and Indigenous peoples. Here is a second parallel with Warren's arguments about Mesolithic Irish foragers. Warren argues that Irish hunter-gatherers should matter to us, even if their population densities were low and their buildings and artifact assemblages comparatively sparse. Mesolithic Irish foragers matter because they occupied and modified the land over a long period of time; because they exemplified a hunter-gatherer existence for which we have few ethnographic parallels; and because they influenced the people who came after them. And, of course, because they were people who lead meaningful lives within their own social worlds.

That people in the past matter is perhaps the greatest lesson of modern archaeology. The world history that I learned as a child, focused on "the Rise of Western Civilization" and similar residual nonsense of the colonial era, pretended that the lives of most Africans, Asians, and Indigenous Americans, Australians, Pacific Islanders did not matter as much as did the lives of Aristotle, Alexander the Great, Charlemagne, and King Henry VIII. The act of considering all human lives – over the past 200,000 or more years of human history and across the globe – as having been important is an emancipatory act, for it reveals that human possibility transcends the limits of the modern world that we generally accept as normal and inevitable.

The Scale Matters workshop was framed around the apparently contradictory conclusions by Bird-David (2017) that hunter-gatherers live in nano-scale societies, and by Bird et al. (2019), that hunter-gatherers have large-scale social networks. Workshop attendees soon came to realize that these arguments were not contradictory. This is demonstrated by Warren's descriptions of the 2020 pandemic lockdown, and the Mesolithic Irish archaeological record. Professor Warren's social life during the pandemic was nanoscale in that he cohabited with the same few human and non-human persons daily. But via telecommunications technologies he exchanged advice, stories, and text with a large range of non-kin, many of whom probably became kin through these exchanges. Likewise, Warren emphasizes that Mesolithic Irish foragers living in tents composed small social worlds of genetic, fictive, and non-human kin. But using the Mesolithic equivalent of Warren's telecommunications technology – long distance trade of stone tools and other raw materials – these scattered small worlds formed a larger-scale social world that spanned the Irish island.

Authors' biographies

Nurit Bird-David is Professor Emerita for Cultural Anthropology at the University of Haifa, Israel. She received her PhD in anthropology from Cambridge University and has carried out the London School of Economics and Political Science and has carried out long-term ethnographic field research in South India. She is "Us, Relatives: Scaling and Plural Life in a Forager World" (University of California Press 2017). Email: nurit.birddavid@gmail.com

Brian F. Codding is Associate Professor of Anthropology and Director of the Archaeological Center at the University of Utah. He received his PhD in anthropology from Stanford University. His research leverages theory from behavioral ecology to examine human-environmental dynamics across ethnographic, ethnological, historical, and archaeological contexts. Email: brian.codding@anthro.utah.edu.

Kasey E. Cole is a PhD. student in Anthropology and Assistant Director of the Archaeological Center at the University of Utah. She received her masters degree in anthropology from California State University, Chico. Her research examines how climate-driven variation in resource distributions influences the formation of social boundaries among hunter-gatherer societies across historical and archaeological contexts. kasey.cole@utah.edu.

Charlotte Damm is Professor of Archaeology at the Arctic University of Norway. She received her PhD from the University of Cambridge and her research has mainly focussed on northern hunter-gatherers. Email: charlotte.damm@uit.no

Stephan Henn is a PhD candidate in anthropology at the University of Cologne. He has worked in the project "Anthropological Models for a Reconstruction of the First African Frontier" that formed part of a Collaborative Research Centre "Our Way to Europe: Culture-Enviroment Interaction and Human Mobility in the Late Quaternary". He is specializing on the connection between ethnography and agent-based modelling. Email: stephan.henn@uni-koeln.de

Robert Kelly is Professor of Anthropology at the University of Wyoming, USA. He received his PhD in anthropology from the University of Michigan, and has been involved with archaeological research in the western US for nearly half a century. He has conducted ethnographic research in Madagascar, and is author of The Lifeways of Hunter-Gatherers (Cambridge 2013) and The Fifth Beginning (U. California Press, 2016). Email: RLKELLY@uwyo.edu.

Andreas Maier is Junior Professor for the Environment and Cultural Evolution of Early Modern Humans at the University of Cologne. He received his PhD in prehistoric archaeology from the University of Cologne, conducted research on palaeodemography and carried out archaeological field research in Central and Eastern Europe. He is author of "The Central European Magdalenian. Regional Diversity and Internal Variability" (Springer 2015). Email: a.maier@uni-koeln.de

Elspeth Ready is a senior researcher in the Department of Human Behavior, Ecology and Culture at the Max Planck Institute for Evolutionary Anthropology in Leipzig, Germany. She received her PhD in Anthropology from Stanford University and conducts ethnographic research with Inuit in Nunavik as well as archaeological research focused on methods and theory for understanding human subsistence ecology in the past. Email: elspeth_ready@eva.mpg.de

Isabell Schmidt conducted research on Pleistocene Population Dynamics within the framework of the CRC 806. She received her PhD in prehistoric archaeology from the University of Cologne on "Solutrean Points of the Iberian Peninsula – Tool making and using behaviour of hunter-gatherers during the Last Glacial Maximum" (BAR 2015). She carried out archaeological field research in Southern Africa and western Europe on Middle Stone Age and Middle and Upper Palaeolithic sites. Email: isabell.schmidt@uni-koeln.de

Bram Tucker is an Associate Professor of Anthropology at the University of Georgia. He received his PhD in anthropology from the University of North Carolina at Chapel Hill. Bram is an ecological, economic, and evolutionary anthropologist who studies household livelihood strategies and the cognitive bases of subsistence decision-making, with emphasis on risk, time, causality, and culture. He has conducted ethnographic research in southwestern Madagascar since 1996.

Graeme Warren is a Professor in the School of Archaeology, University College Dublin. He is a specialist in the archaeology of hunter-gatherers. He is the author of "Hunter-Gatherer Ireland: making connections in an island world" (Oxbow 2022). Email: graeme.warren@ucd.ie; Twitter: @graememwarren

Thomas Widlok is Professor for Cultural Anthropology of Africa at the University of Cologne. He received his PhD in anthropology from the London School of Economics and Political Science and has carried out long-term ethnographic field research in Australia and southern Africa. He is author of "Anthropology and the Economy of Sharing" (Routledge 2017). Email: thomas.widlok@uni-koeln.de

Kurt M. Wilson is a PhD. candidate in Anthropology and Research Associate with the Archaeological Center at the University of Utah. He received his masters degree in anthropology from Iowa State University. His work couples agent-based and statistical modeling with ethnological and archaeological data to examine the ecological foundations of human social organization, including territoriality and inequality. kurt.wilson@utah.edu.

Andreas Zimmermann is Emeritus Professor for Neolithic Archaeology at the Institute for Prehistory, University of Cologne. His research interests are population dynamics, diachronic comparisons, landscape and economical archaeology as well as quantitative methods. a.zimmermann@uni-koeln.de.

Cultural Studies

Gabriele Klein
Pina Bausch's Dance Theater
Company, Artistic Practices and Reception

2020, 440 p., pb., col. ill.
29,99 € (DE), 978-3-8376-5055-6
E-Book:
PDF: 29,99 € (DE), ISBN 978-3-8394-5055-0

Ingrid Hoelzl, Rémi Marie
Common Image
Towards a Larger Than Human Communism

2021, 156 p., pb., ill.
29,50 € (DE), 978-3-8376-5939-9
E-Book:
PDF: 26,99 € (DE), ISBN 978-3-8394-5939-3

Elisa Ganivet
Border Wall Aesthetics
Artworks in Border Spaces

2019, 250 p., hardcover, ill.
79,99 € (DE), 978-3-8376-4777-8
E-Book:
PDF: 79,99 € (DE), ISBN 978-3-8394-4777-2

All print, e-book and open access versions of the titles in our list are available in our online shop www.transcript-publishing.com

Cultural Studies

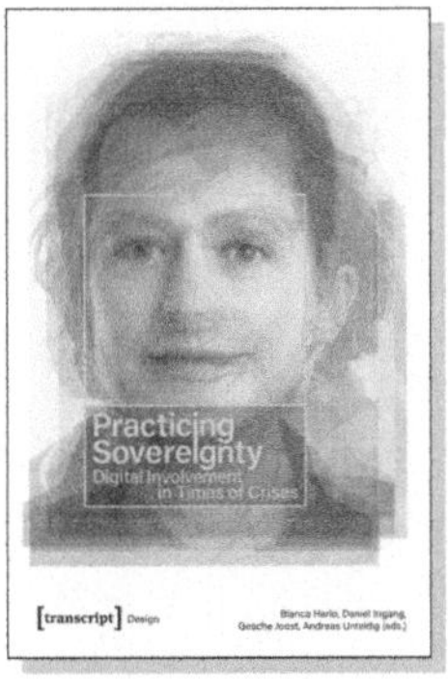

Bianca Herlo, Daniel Irrgang, Gesche Joost, Andreas Unteidig (eds.)
Practicing Sovereignty
Digital Involvement in Times of Crises

January 2022, 430 p., pb., col. ill.
35,00 € (DE), 978-3-8376-5760-9
E-Book: available as free open access publication
PDF: ISBN 978-3-8394-5760-3

Tatiana Bazzichelli (ed.)
Whistleblowing for Change
Exposing Systems of Power and Injustice

2021, 376 p., pb., ill.
29,50 € (DE), 978-3-8376-5793-7
E-Book: available as free open access publication
PDF: ISBN 978-3-8394-5793-1
ISBN 978-3-7328-5793-7

Virginie Roy, Katharina Voigt (eds.)
Dimensions. Journal of Architectural Knowledge
Vol. 1, No. 2/2021:
Spatial Dimensions of Moving Experience

2021, 228 p., pb., ill.
39,00 € (DE), 978-3-8376-5831-6
E-Book: available as free open access publication
PDF: ISBN 978-3-8394-5831-0

All print, e-book and open access versions of the titles in our list are available in our online shop www.transcript-publishing.com